The Identification of Furniture Woods

Arthur Koehler

SOTHIS PRESS

ISBN 978-1-0365-1753-3

This edition is published by
Sothis Press © 2023. The company
is registered in England and Wales.
The Company Number is 14900389.

SOTHIS PRESS

Sothis Press aims to completely reimagine the public domain book with an emphasis on universal access to captivatingly designed print editions. We believe that technology will enable every single book ever published to be perpetually accessible in beautiful print form. Sothis is beauty in ubiquity.

www.sothis.press
1 Waterhouse Square, London, EC1N 2ST

UNITED STATES DEPARTMENT OF AGRICULTURE

MISCELLANEOUS CIRCULAR NO. 66

WASHINGTON, D. C. NOVEMBER, 1926

THE IDENTIFICATION OF FURNITURE WOODS

By Arthur Koehler, *in Charge, Office of Wood Technology, Forest Products Laboratory, Forest Service*

CONTENTS

WOOD FOR FURNITURE

Ever since our ancestors left off nomadic habits and settled down in one place long enough to want furniture, it has been made chiefly out of one kind of material—wood. Whether the furniture maker was Egyptian, Greek, or Roman, whether he was hacking out a crude piece of usable equipment or striving to express an ideal of beauty in the form and quality of his creation, he preferred nearly always to work in wood. Styles have changed; reeds, fiber, and even metal have been used to some extent, but wood has been and very likely will continue to be the chief material out of which furniture is made.

The reason for the almost universal use of wood for furniture is that it combines the advantages of being fairly abundant and of having certain properties which make it especially suitable for the purpose. The principal properties are:

Wood is easily worked.

It is easily fastened together with nails, screws, dowels, or glue.

It is or may be made highly ornamental.

It is a poor conductor of heat and therefore does not feel very cold or very hot when of a temperature differing considerably from the body temperature.

It is comparatively noiseless under impact.

It is easily repaired.

Many species of wood have such a combination of weight and strength that they can be fashioned into articles of agreeable proportions and of sufficient strength without excessive weight.

SELECTION OF FURNITURE WOOD

Wood for the exposed parts of most kinds of furniture must be sufficiently hard to resist indentation in ordinary usage. Many woods are strong enough and possess other qualities which would fit them for use in exposed parts of household furniture but are barred for lack of the requisite hardness. Two good examples of such woods are butternut and Spanish cedar. On the other hand, the wood must not be too hard, or it will be so difficult to work as to increase materially the cost of manufacture.

Wood which has the requisite hardness also is sufficiently strong not to break in handling or in use, unless weakened by defects. On the other hand numerous woods, like basswood, cottonwood, chestnut, or willow, are not only too soft but too weak (unless used in comparatively large sizes) for the substantial parts of high-grade furniture. Even the stronger woods, such as oak and walnut, fail occasionally in use because of cross grain, decay, or too deep carving or mortising.

Good appearance is required in wood used for most kinds of furniture. Many species which otherwise would be suitable for high-class furniture have not a sufficiently attractive figure or color to make them desirable. Maple, for example, is not used as much as its strength and hardness would warrant because it usually possesses no distinctive figure.

Comparative freedom from warping and from excessive shrinking and swelling is essential. Although proper seasoning methods and reasonable care of furniture reduce to such an extent the troubles due to warping that even woods with a reputation for warping excessively may be and are used for high-grade furniture, it is always safer to construct furniture from species that will hold their shape well. The enviable reputations of mahogany and walnut are due in part to this characteristic, and the usual preference of quarter-sawed to plain-sawed lumber is based on the common experience that the quarter-sawed warps less. Changes in the dimensions of wood can not, however, be avoided entirely. As a general rule, the heavier and harder species shrink more than the lighter and softer ones.

To season satisfactorily, work easily, glue without much difficulty, and finish well also are important from the manufacturer's standpoint.

Proper seasoning does not necessarily mean that the lumber shall stand in piles air drying for a given number of years, but simply that it shall be dried to a proper and uniform moisture content and be free from seasoning defects and internal stresses. In the seasoning of lumber the modern manufacturer, with his closely-regulated dry kilns, has a distinct advantage over the old-time cabinetmaker, who had no such equipment and therefore often turned out a product which shrank, warped, or checked in use.

Resistance to decay is not an important factor in the selection of wood for household furniture, since practically all furniture in use is too dry to be subject to decay. In fact, maple, birch, sap gum, and other woods which rank low in durability in damp locations are used extensively and with good results in household furniture. If, however, such woods are not piled openly and otherwise taken care of properly before they are fully seasoned, they may become sufficiently infected with decay-producing fungi to be seriously weakened and made brittle so that they can not be finished smoothly.

Porch and lawn furniture is occasionally, and indoor furniture rarely, attacked by the powder-post beetle when made of sapwood or ash, hickory, elm, oak, or other hardwoods. The use of heartwood only or keeping the furniture well painted will obviate attack by this insect.

AMOUNT AND KINDS OF WOOD USED FOR FURNITURE

Throughout its history the United States has been fortunate in having a goodly supply of various species of wood suitable for furniture manufacture. Except mahogany, comparatively little wood has been imported for this purpose. Although our hardwood forests have been disappearing rapidly during the past century, considerable stands of oak, maple, red gum, walnut, and other excellent species remain. With better forestry practice on public and private holdings and less waste in manufacturing, it will still be possible to keep up a supply of hardwoods sufficient to meet a large part of the demand for furniture. Fortunately, timber can be grown in never-ending quantities and does not need to be "mined" from a limited original supply. The pride of Americans in their homes and country should find expression for many years to come in furniture of American material, American design, and American manufacture.

The total amount of lumber used annually in the United States in the manufacture of furniture of all kinds is between 1¼ and 1½ billion board feet. In quantity of lumber used, furniture manufacturing ranks among the foremost industries, being surpassed only by the planing-mill and box-manufacturing industries and equaled by railway-car construction.

Hardwoods, or woods from broad-leaved trees, predominate in the manufacture of furniture. Softwoods, or woods from trees with needlelike or scalelike leaves, commonly called "conifers," are at present used only to a very limited extent. As the supply of conifers is far more abundant, however, than that of hardwoods, the conifers undoubtedly will be used more and more, especially for hidden parts such as interiors and the cores of veneered panels.

The principal woods used in the United States in the manufacture of exposed parts of household furniture, not including kitchen, porch, and lawn furniture, are those described on pages 27 to 71 of this circular—with oak, red gum, maple, black walnut, birch, and mahogany leading. The exact order of importance, however, varies with the styles. This circular does not cover those species of wood which are used almost exclusively in hidden parts of furniture or only in kitchen furniture—such as yellow poplar, basswood, cottonwood, pine, fir, and spruce; nor those used in this country for exteriors to a very limited extent only, such as holly, cypress, satinwood, teak, and redwood burl. Complete cataloguing would have made the circular too cumbersome and the identification key too involved to be readily used in the identification of the more important species.

THE STRUCTURE OF WOODS

The identification of woods is not a simple matter, because there are many kinds and each kind varies more or less in appearance and properties within itself. For example, some mahogany lumber is light colored and other lumber of the same species is dark colored. Some pieces of mahogany are twice as heavy as others of the same size. Furthermore, there are often puzzling superficial *resemblances* between unrelated species; as, for example, between birch and maple and between cherry and mahogany.

Lumbermen, manufacturers, and dealers in furniture usually recognize the more common woods by general appearances. Though this method is fairly accurate for a limited number of woods, it does not lend itself to description, nor can it be used very well for the instruction of others or in settling disputes. For such purposes it is necessary to point out specific differences.

The most pronounced and reliable specific differences in the appearances of woods are found in their cellular structure. For this reason the key for identification, pages 71 to 76, is based to a large extent on structure. Color, odor, weight, and hardness help in identifying woods, but as a rule such qualities are too variable to be used singly in distinguishing a large number of woods. This drawback applies particularly to color, which not only varies in the natural wood but is modified by treatments, paints, stains, and decay.

The chief structural characteristics of the hardwoods[1] used for furniture are the pores, the annual rings, and the rays.

Hardwoods are made up principally of vertically elongated cells, which constitute the fibers and pores,[2] and transversely elongated cells aggregated into strips running at right angles to the fibers from the bark toward the center and known as medullary rays or simply rays (pl. 1).

The fibers form the body of the wood. Individually they are too small to be seen without a high-power microscope. Usually they are not significant in identifying wood.

The pores are large enough in some woods to be seen distinctly with the naked eye, appearing as minute holes on smoothly cut transverse, or end, surfaces and as grooves on smooth longitudinal surfaces of the wood. *A* of Plates 10, 15, 18, and 21, for example, show transverse surfaces of such woods, somewhat magnified.[3] In

---

[1] See p. 3 for meaning of hardwoods and softwoods.

[2] Other vertical cells used principally for storing food and known as parenchyma, and comparatively small elongated cells which conduct sap and are known as tracheids are also found in nearly all hardwoods; but, to simplify the text, since it is not important to refer to them specifically, they are not further mentioned in this circular.

[3] Although the descriptions of woods in this circular are based largely on what can be seen without the use of a magnifying glass, the illustrations of end surfaces of woods are magnified seven and one-half diameters, since reproductions of natural-sized photographs of the end surfaces do not bring out the details as clearly as they appear on the actual wood. The enlarged views of end surfaces also show with reasonable clearness the features mentioned in the text as being visible under a lens. The magnification of seven and one-half corresponds to that of ordinary hand lenses. Illustrations of the radial and tangential surfaces are natural size, since the general appearance of these surfaces is more important for purposes of identification than any structural details that might be revealed at moderate magnifications.

Cube of white oak wood magnified about 4.3 diameters. End surface at top; radial, or quarter-sawed, surface at left; tangential, or plain-sawed, surface at right. *AR*, annual rings; *P*, large pores containing tyloses, *SP*, small pores arranged in radial rows; *MR*, large medullary rays. Note very small rays between larger ones other hardwoods the pores are so small that they are not visible without a magnifying lens. *A* of Plates 23 to 30 show such woods. The difference in the size of pores therefore forms one basis for classifying woods. Birch has pores large enough to be made out with the naked eye, but only under very favorable conditions; hence in the key for identification, pages 71 to 76, it is classified both as having pores visible and as having pores not visible without a magnifying glass.

In certain hardwoods the pores, after they have conducted sap for a number of years, become clogged with bubblelike ingrowths from neighboring cells. These ingrowths, which are known as tyloses, usually appear as a glistening or iridescent substance in the pores, especially when viewed under a lens. In certain other hardwoods the pores become partly or entirely filled with a colored gum, while in still others they remain open. The character of the pore contents is a reliable guide in classifying woods.

As a rule, in woods grown in climates where a season of growth is followed by a season when growth is as a standstill, the pores and fibers formed at the beginning of each season's growth are larger or have thinner walls than those formed toward the end of the season, thus making a sharp line of demarcation between the growth of one season and that of the next. The seasonal growth layers are called "annual rings." Many tropical woods grow all year long and show no well-defined annual rings. The absence of annual rings in khaya, tanguile, red lauaan, and almon helps in distinguishing these woods from true mahogany. Compare *A* of Plates 18, 19, and 20.

In some woods the pores formed at the beginning of each annual ring are so much larger than those formed later in the year's growth that well-defined porous zones are formed, as illustrated in *A* of Plates 10 to 15. Hardwoods of this type of structure are classed as "ring-porous." In many other woods there is no abrupt difference in the sizes of the pores formed early in the year's growth and those formed later (*A* of pls. 16 to 30). Hardwoods of this type are classed as "diffuse-porous." Seasonal variation in the size of the pores and the absence of such variation are therefore characteristics useful in classifying woods.

Another characteristic useful in identification is the arrangement of the pores in different kinds of hardwoods. In some species, groups or lines of pores extend across the outer portion of each annual ring—that is, radially as seen on the end surface. This is the case in oak and chestnut (*A* of pls. 10, 11, and 12). In other woods the pores are arranged in more or less wavy lines extending along the annual rings on the end surface, as in elm and white ash (*A* of pls. 13, 14, and 15). In many woods there is no definite arrangement of the pores, as in the walnuts, birches, and maples.

The rays in all woods are very narrow strips of cells extending radially across the annual rings, but there is a noticeable difference in their width. On end surfaces the rays are distinctly visible in some woods without magnification, and in others not at all. In oak, for example, which has by far the largest rays of our native trees, they appear as conspicuous lines on end surfaces. *A* of Plates 10 and 11 show them magnified seven and one-half times. Vertically in the tree the rays may be as high as 4 inches in oak, but in all other native species they are less than one-fourth inch in height. *C* of Plates 10 and 11 show radial views of the large rays in oak. Next to oak in size of rays comes sycamore, then beech. Sometimes the rays contain coloring matter and appear as darker flakes when cut lengthwise. Among furniture woods this holds especially for tanguile, red lauaan, almon, maple, sycamore, and beech. The size and color of the rays therefore afford additional characteristics useful in the identification of woods.

Sometimes the rays are confused with the annual rings on end surfaces of lumber. The two can be differentiated by the fact that the rays are relatively thin, usually straight lines converging slightly toward one side or edge, whereas the annual rings are comparatively broad curved zones.

The rays in true mahogany often are in tiers or stories in the tree and appear in minute rows or ripplelike marks running across the grain on plain-sawed surfaces. This arrangement of the rays can be seen in certain parts of Plate 18, *B*. Similar but more obscure "ripple marks" (usually not visible without a magnifying glass) are found in rosewood. Only strictly plain-sawed surfaces should be examined for the presence of "ripple marks." In mahogany there are about 50 per inch and in rosewood close to 150. With rare exceptions over small plain-sawed areas this arrangement of the rays is not found in any other of the woods described in this circular.

Tree trunks, as a rule, have a light-colored outer layer of wood known as sapwood and a darker core known as heartwood, although exceptions occur in which the heartwood is practically as light colored as the sapwood. Mention of the color of wood in this circular refers to the color of the heartwood, unless otherwise stated, since heartwood lumber is generally used in making furniture. There is no essential difference, however, between sapwood and heartwood for this purpose, except in natural color and in the way they take stains. Differences between them in strength, hardness, and durability are insignificant under the conditions in which household furniture is ordinarily used. The comparatively thick sapwood developed by some species, notably red gum, maple, birch, and ash, is used extensively in the manufacture of furniture.

METHODS OF CUTTING LUMBER FROM LOGS

Because wood has annual rings and medullary rays, it can be cut longitudinally in two distinct ways: (1) At right angles to the rays, or tangentially to the growth rings, producing what is known as tangentially cut, plain-sawed, or flat-grain lumber; and (2) parallel to the rays and across the rings, producing what is known as radially cut, quarter-sawed [4] or edge-grain lumber (fig. 1). In the case of hardwoods, "plain-sawed" and "quarter-sawed" are terms most commonly used.

Plain-sawed lumber is usually cheaper than the quarter-sawed, because it can be cut with less waste; and in some kinds of wood, such as ash, elm, chestnut, and all coniferous species, it has a better figure.

[4] The term "quarter-sawing" is derived from the early practice of cutting a log into quarters and sawing each into boards as nearly parallel to the faces as practical.

Quarter-sawed lumber shrinks less in width and is less subject to checking, twisting, and cupping; and in some woods—oak and mahogany, for example—it usually has a better appearance. Figure 2 shows a very common method of quarter-sawing lumber. A number of modifications are also in use. Any method of quarter-sawing necessarily produces considerable lumber which is cut at more or less of an angle with the rays and hence is not perfectly quarter-sawed. Because such methods require more time and cause more waste, and also because quarter-sawed lumber of many woods

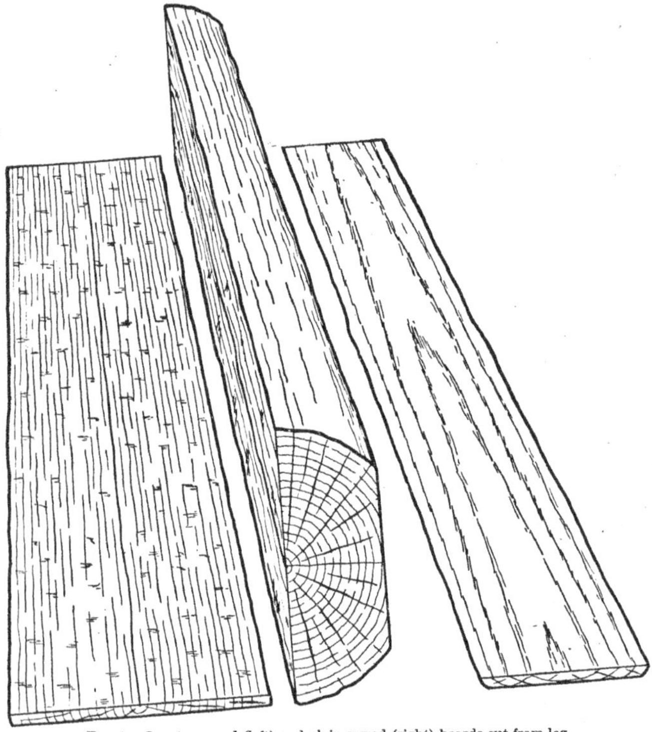

FIG. 1.—Quarter-sawed (left) and plain-sawed (right) boards cut from log

is considered superior to the plain-sawed in beauty as well as in quality, quarter-sawed material usually commands the higher price.

VENEERS

Veneers are thin sheets of wood which in the manufacture of furniture are usually glued together or on thicker lumber, though occasionally single sheets of veneer are used, as for the backs of bureaus and similar pieces, backs of mirrors, and for drawer bottoms. Veneers are either cut in straight sheets from a squared timber or flitch or peeled off in a continuous layer from a rotating log, that from the latter process being known as rotary veneer. Straight sheets are cut either with a saw or with a knife, the products being known respectively as sawed and sliced veneer.

Sawed and sliced veneers have the advantage that they can be cut from a flitch in any direction. Sawed veneer usually is cut radially so as to produce the types of figure that are obtained by quarter-sawing, such as the "silver grain" in oak, but sliced veneer is cut either radially or tangentially or along intermediate planes.

Rotary veneer has the advantage that it can be produced cheaply and in large sheets and that its growth-ring figure is continuous, whereas the figure in plain-sawed lumber merges toward both edges of a face into a form intermediate between a strictly plain-sawed and a quarter-sawed figure. Compare *A* and *B* of Plate 2.

FIGURE IN WOOD

Figure in wood is the pattern formed by irregular infiltrations of coloring matter; by the annual rings and the medullary rays;

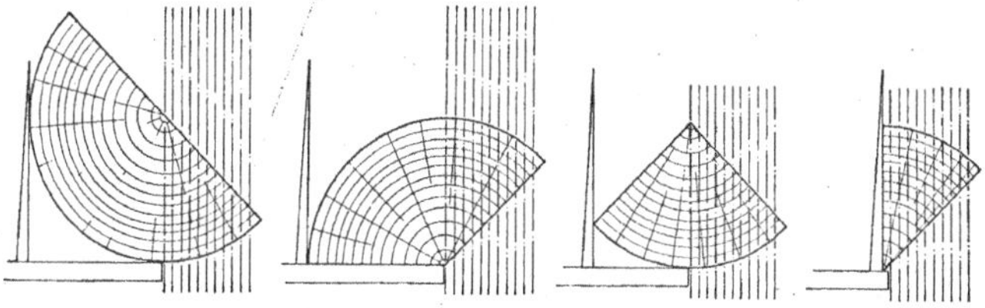

FIG. 2.—Common method of quarter-sawing lumber. The dash-and-dot lines represent successive saw cuts

and by cross grain, wavy grain, burls, knots, or other distortion of the normal course of the fibers.

Natural differences in color are comparatively rare in furniture woods, except as the heartwood differs from the sapwood. Circassian walnut, rosewood, some red gum, and (to a minor extent) some black walnut contain darker streaks which give the wood a distinctive figure. In some hardwoods the figure often is made more prominent by the use of dark or, occasionally, light-colored fillers. In conifers unequal absorption of stains produces a like effect.

In many woods in which the color is fairly uniform throughout, some parts look darker than others because of changes in the direction in which the fibers extend, causing light to be reflected differently from different areas. With changes in the angle of vision or of the source of light, the reflection also changes, making the portions previously dark appear bright, and vice versa. The greater the natural luster of the wood, the more pronounced are such differences in reflection. This is the underlying cause of the beauty of mahogany and many other figured woods; and by virtue of this play of light and shade certain kinds of natural figure can very easily be distinguished from stained or painted imitations.

All woods have some figure, but in many it is so obscure or uninteresting as to receive no consideration. It is because some woods

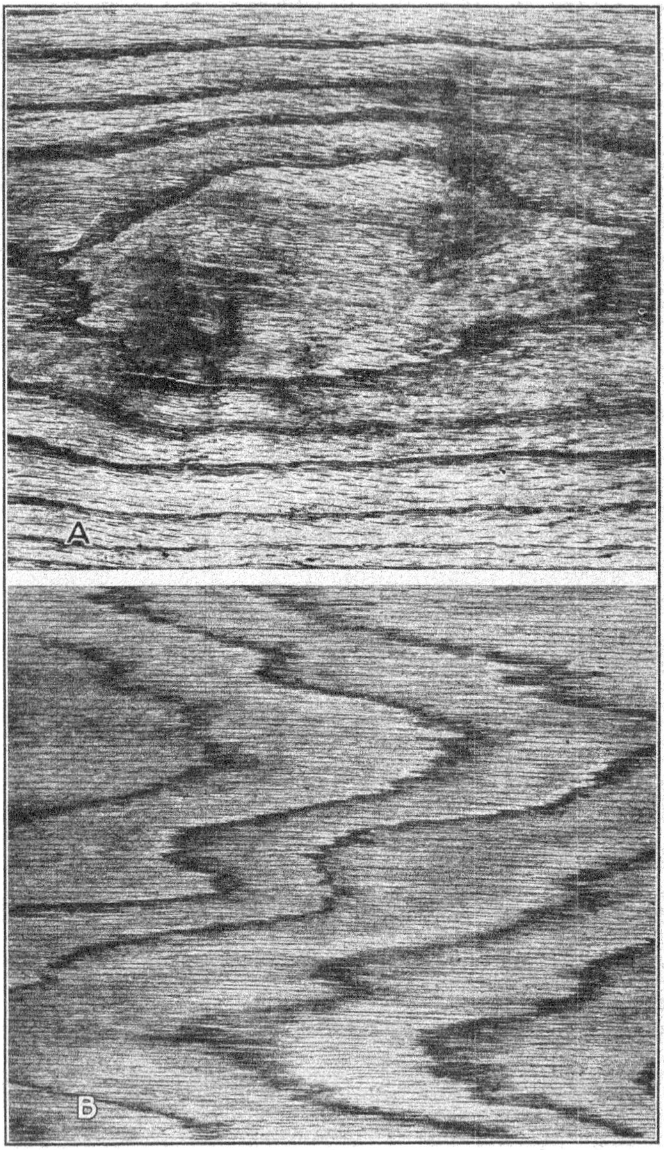

A. Growth-ring figure in plain-sawed oak. *B.* Growth-ring figure in rotary-cut oak veneer. (About one-half natural size)

have a distinctly variegated appearance that they are considered superior and are used in the higher grades of furniture. A uniform color, no matter how rich its tone, does not give so pleasing an effect as the variations in color, lights and shadows, curves and stripes, which form the figure.

No two pieces of wood show the same figure. In this way the natural pattern of wood differs from artificial patterns such as are sometimes painted on wood or metal or incorporated into fabrics of various kinds, except when such patterns are executed individually and at great expense. Thus a piece of furniture for which the wood has been carefully selected and properly manufactured and finished is an exhibition not only of the mechanical skill of man but also of beauties of nature which can be revealed to best advantage only by the hand of the artisan.

The following are the principal kinds of figure recognized in furniture woods:

GROWTH RING

Wherever the seasonal layers of growth are differentiated, as is the case in most woods grown in temperate climates, these layers, or annual rings, form a pattern when cut along any plane. This pattern consists of stripes on quarter-sawed surfaces and of stripes, parabolas, and ellipses (all more or less irregular) on plain-sawed surfaces, as shown in Plate 2, *A*, and Plate 3, *A* and *B*. In lumber cut from crooked or bulged logs the figure often assumes peculiar shapes. It is most strikingly displayed in rotary-cut veneer.

Growth-ring figure is very pronounced in oak, ash, chestnut, and elm, and moderately so in black walnut, butternut, birch, and pecan.

SILVER GRAIN

Silver grain is a name applied to the large medullary rays which appear as "flakes" or "flecks" on the faces of quarter-sawed oak lumber, as shown on the front cover of this circular. When lumber or veneer is cut almost parallel radially with the rays and slightly diagonally longitudinally, the portions of the rays which show will run at an angle across the face of the board. When adjacent pieces so cut are matched so that the angle is reversed, a "herring bone" figure is formed.

PIGMENT FIGURE

The term pigment figure, although not common in the furniture trade, is here used to include such figures as are due to differences in the color of the wood, exclusive of such contrasts as appear in the growth-ring figure. The iregular dark streaks in Circassian walnut, rosewood, and red gum, which give "life" to these woods, come under this heading (pl. 4, *A* and *B*.) Streaked red gum is usually spoken of as "figured red gum" without any particular name being given to the figure.

STRIPE OR RIBBON

This type of figure consists of alternating lighter and darker stripes, running more or less the length of a board and varying from less than one-fourth of an inch to more than 1½ inches in width.

A. Growth-ring figure in plain-sawed elm. *B*. Growth-ring figure in plain-sawed white ash. (About one-half natural size)

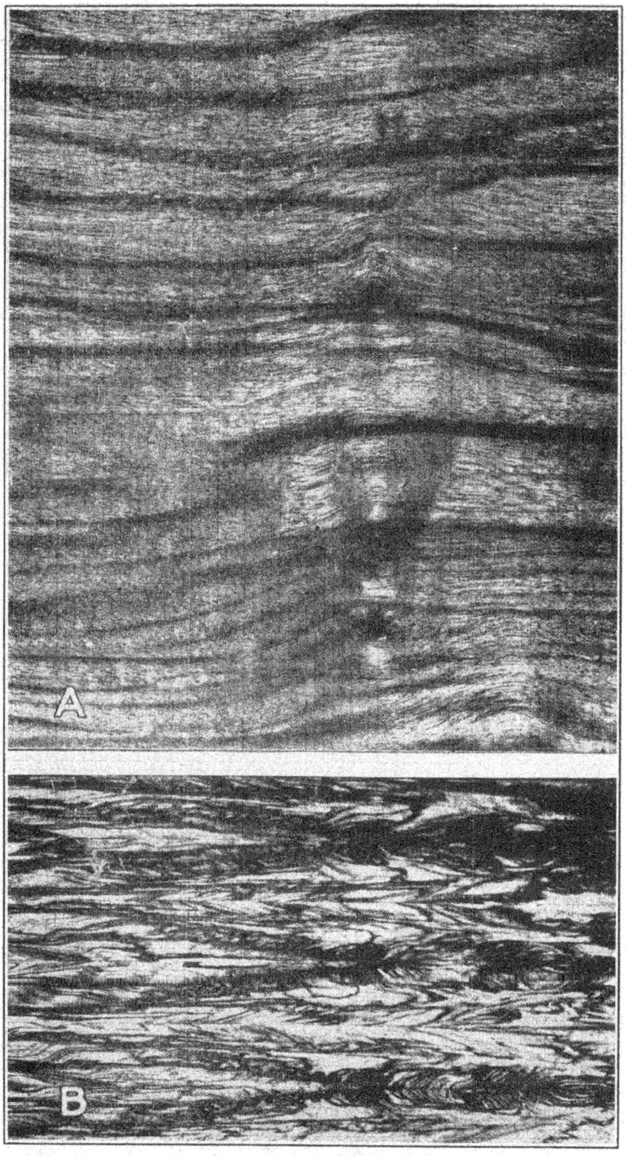

A, Pigment figure in Circassian walnut. (About two-thirds natural size.) *B*, Pigment figure in matched red gum veneer. (On panel of "one-panel" door)

It is pronounced only in quarter-sawed or nearly quartered lumber. It usually is due to differences in the reflection of light from adjacent layers of wood cut from trees with interlocked grain—that is, trees in which the slope of the fiber alternates between a right-handed and a left-handed spiral around the tree in succeeding periods of growth, each period usually extending over a number of years. This configuration is illustrated in text Figure 3. Lumber cut radially from such trees contains stripes in which the grain runs alternately inward and outward (pl. 5, *A*), and hence causes

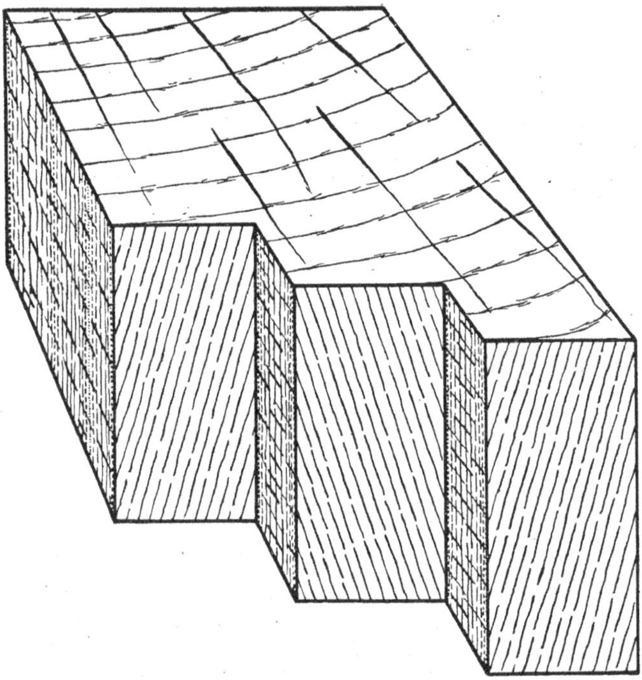

FIG. 3.—Sectional view of block of wood with interlocked grain

differences in the reflection of light. Stripe of this character is common in true mahogany, khaya, tanguile, red lauaan, almon, red gum, and tupelo gum, and in some other woods, principally from the Tropics. In the gums it is not so pronounced as in the other woods mentioned, because the former have not a high natural luster.

Stripe figure is sometimes formed by the annual rings in quarter-sawed lumber, being especially noticeable when the yearly layers of growth are wide. Plate 5, *B*, shows such stripe in quartered black walnut. Wood having fairly straight and regular zones defined by actual differences in color is another case referred to as stripe figure.

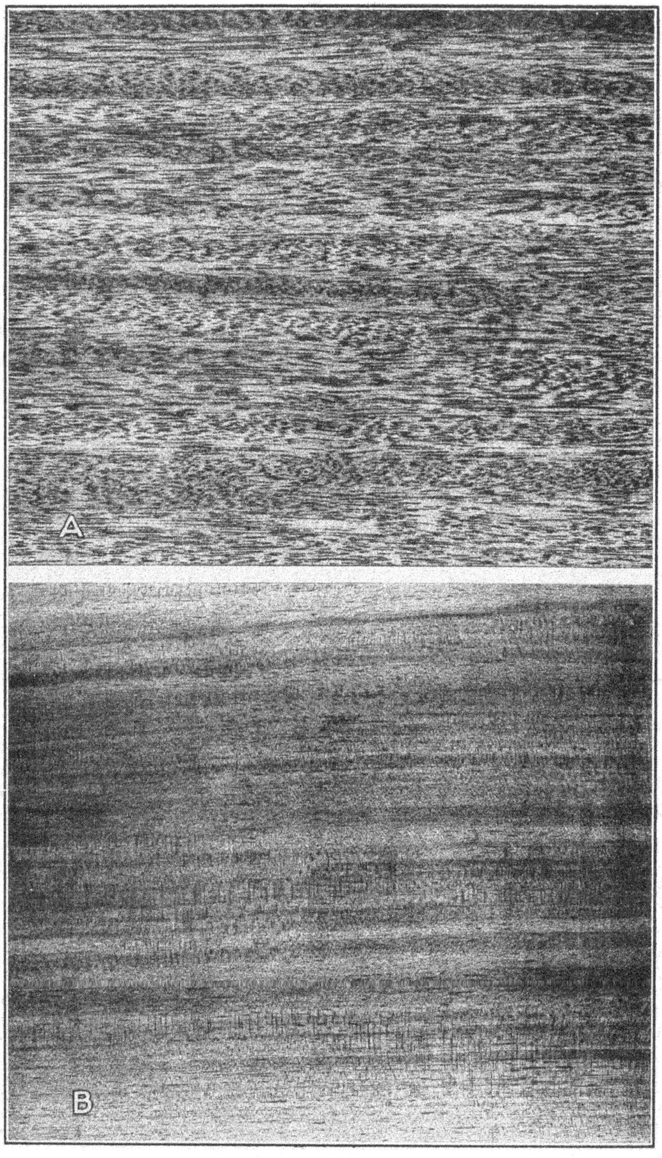

A. Stripe figure in tanguile. *B.* Stripe figure in black walnut. (About one-half natural size)

BROKEN STRIPE

Broken stripe is a modification of the stripe figure, the markings tapering out after extending for a length of about 1 foot to several feet. It is due to undulations in the spiral course of the fibers around the tree, which produce changes in the angle of the fibers. Among furniture woods this type of figure is found principally in mahogany. It is illustrated in Plate 6, *A*.

ROE

Roe is a term applied to short broken stripe, a type of figure occurring chiefly in mahogany.

MOTTLE

Mottle is a variegated pattern which consists principally of local, irregular, wavy configurations of the fibers extending for short distances across the face of boards or veneers. It may be seen on quarter-sawed surfaces, where it often is combined with stripe, as in mahogany (pl. 6, *B*), or it may occur on tangential surfaces, especially of butt veneers, as in walnut (see back cover).

FIDDLE-BACK

Fine regular waves in the grain of the wood produce a figure which is known as fiddle-back because wood, especially maple, with this kind of figure has long been used for the construction of violins. It is illustrated in Plate 7, *A*. Fiddle-back is found principally in mahogany and maple, but occurs in many other species. It shows to best advantage on truly radially cut surfaces.

RAINDROP

When the waves in the fibers occur singly or in groups with considerable intervals between, producing streaks here and there across the face of the board, the figure is called "raindrop," because it suggests the streaks made by drops striking a surface at a slant (pl. 7, *B*).

FINGER ROLL

Finger roll is a wavy pattern in which the waves are about the size of one's finger. Like all true wavy grain it is most pronounced on strictly radially cut lumber.

Figures which extend across the grain, as mottle, fiddle-back, raindrop, and finger roll, are often spoken of as "cross figure."

CURLY AND LANDSCAPE FIGURE

In certain portions of some trees and throughout others the fibers are distorted into irregular convolutions. Wood with such grain is referred to as having a "curly" figure. Sometimes the figure assumes fantastic forms to which the name "landscape figure" has been applied. Among furniture woods curly grain occurs most frequently in maple but may be found in other species.

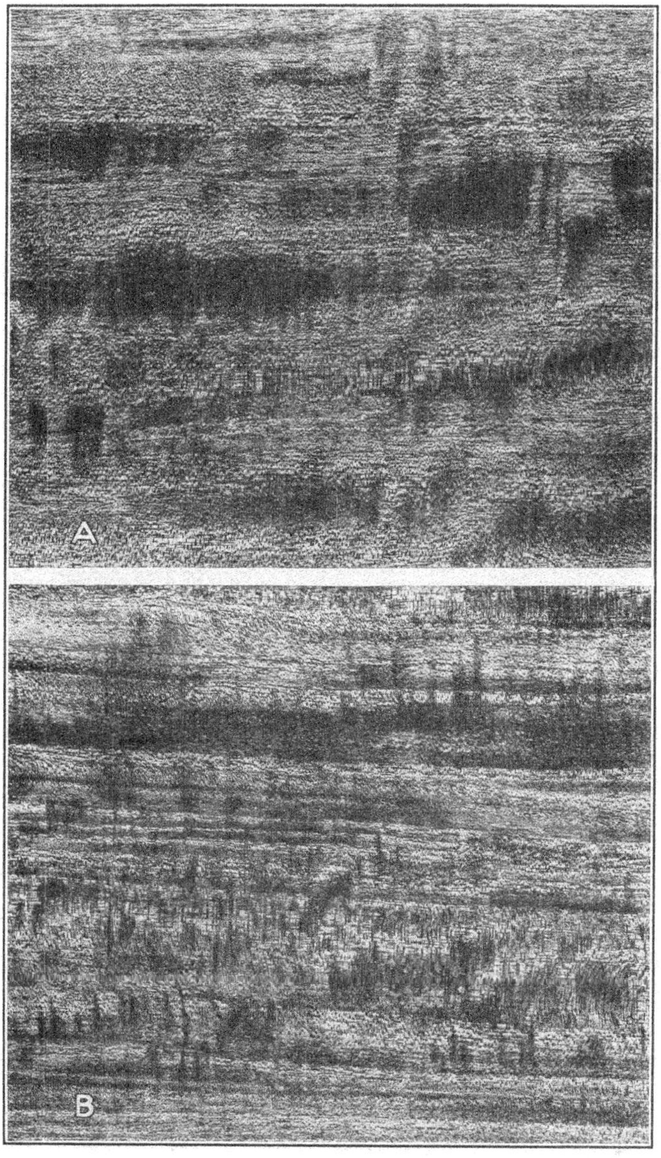

A. Broken stripe in quarter-sawed mahogany. *B*. Mottle figure in quarter-sawed mahogany. (About one-half natural size)

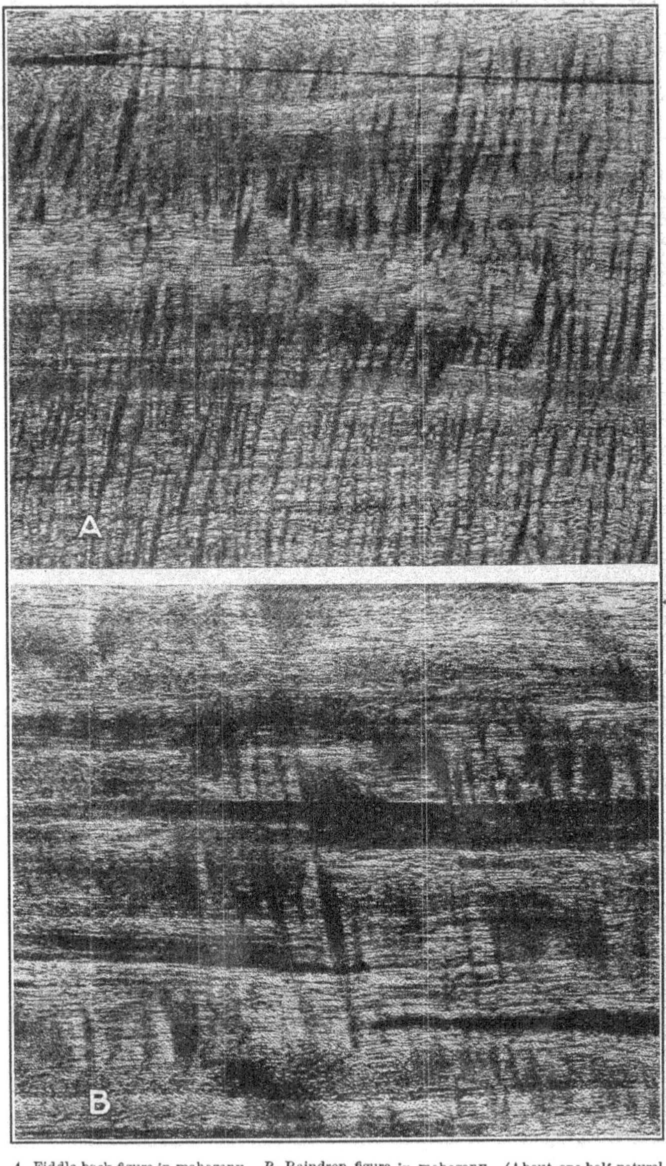

A. Fiddle-back figure in mahogany. *B*. Raindrop figure in mahogany. (About one-half natural size)

STUMP OR BUTT WOOD

At the junction of the larger roots with the stem of a tree and to a less extent where the branches join the stem, the fibers are considerably distorted, producing various kinds of cross figure, particularly mottle and curly grain. Veneers cut from such portions of the tree are highly figured and command a high price. Black walnut stumps find a ready market at veneer plants for this reason. Such veneers are usually matched so as to produce a panel with a symmetrical patern such as is shown on the back cover of this circular.

CROTCH

At forks in the tree trunk or where large branches join the trunk, the fibers run in different directions in closely adjoining portions, often giving lumber cut from such parts of a tree a pattern resembling a cluster of plumes (pl. 8, *A*). This is called "crotch figure," or, more often, "crotch mahogany," "crotch walnut," etc., according to the kind of wood.

BLISTER

Blister consists of seeming knoll-like elevations in the wood, as shown in Plate 8, *B*. It is due to an uneven contour of the annual rings, and not to blisters or pockets in the wood as the name might indicate. It occurs in various species and is pronounced only in plain-sawed lumber or rotary-cut veneer.

BURLS

Burls are large wart-like excrescences on tree trunks. They contain the dark piths of a large number of buds. The formation of these buds, which rarely develop, apparently is due to some injury sustained by the tree. Throughout the burl the fibers are very irregularly contorted so that the grain can not be said to run in any particular direction. Burls may occur on almost any species, but walnut, ash, cherry, and redwood burls are among the most highly prized in furniture woods. Plate 9, *A*, shows burl in cherry.

BIRD'S-EYE

Bird's-eye is due to local sharp depressions in the annual rings, accompanied by considerable fiber distortion. Once the depressions are formed, succeeding growth rings follow the same contour for many years. In plain-sawed lumber and rotary veneer the depressions are cut through crosswise and show a series of circlets, portions of annual rings, suggesting, rather remotely, a bird's eye (pl. 9, *B*). Among native commercial woods typical bird's-eye figure is confined almost exclusively to maple, and it occurs in only a small percentage of maple trees. It differs from burl in that burl is due to conical elevations each containing a dark speck, the pith of a bud. Burl also shows greater distortion of fibers throughout. It is not clear what causes the irregularities which give rise to bird's-eye figure.

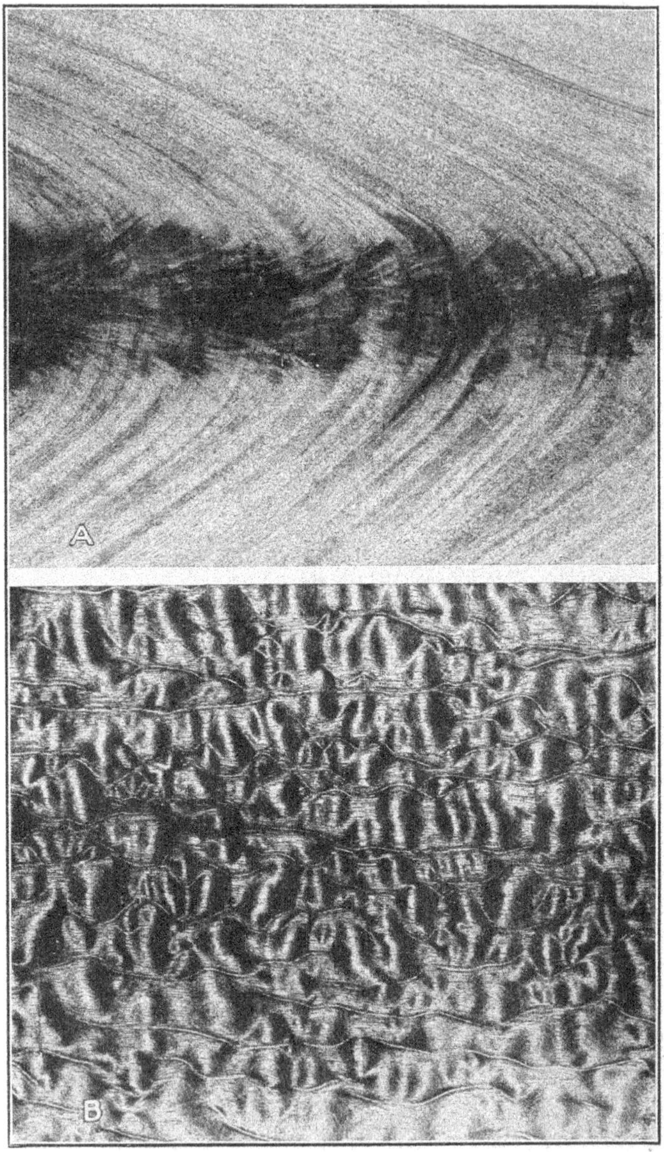

A. Crotch figure in mahogany. *B*. Blister figure in mahogany. (About one-half natural size)

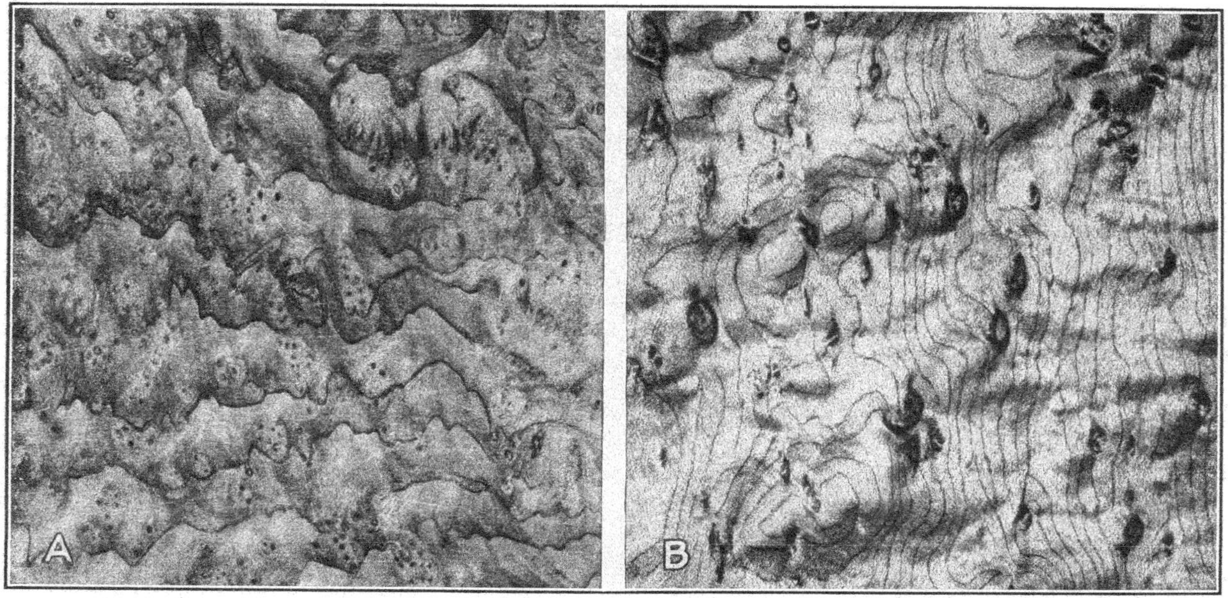

A. Cherry burl. *B.* Bird's-eye maple. (Natural size)

SWIRL

Some trees have barrel-shaped bulges in the trunk in which the fibers are considerably contorted. Veneers cut from such portions show more or less curly grain somewhat resembling burl wood but not containing the dark cores of buds.

KNOTS

At present lumber or veneer containing knots is not popular for use in household furniture other than cedar chests, for which knotty lumber is actually preferred. At one time thin cross sections of limbs with an "oyster shell" figure, due to the eccentric growth of the annual rings of the limbs, were used as decorative veneer Knots, being the bases of limbs, give a similar figure. Highly polished and properly finished knots have a decorative value which might well be made use of to a much greater extent than it is at present. In Europe both the manufacturer and the consumer of furniture realize that certain of the so-called defects and blemishes of lumber not only do not detract from the utility of the product but often enhance its appearance. In colonial furniture, now so highly prized, knotty lumber was freely used for exposed parts in which strength was not essential. This was done purely for decorative purposes, since clear lumber was relatively cheap at that time.

Occasionally other kinds of figure occur in wood. Also the above-mentioned figures are sometimes known under different names. These names, therefore, can not be considered a complete catalogue of the various types of ornamental patterns found in wood. The types here given, however, will be found helpful in getting a better understanding of the structure of wood and in evaluating it as furniture material.

VENEERED VERSUS SOLID CONSTRUCTION

The identification of wood in veneers is sometimes difficult on account of the thinness of the layers and their different orientations in glued-up stock. A brief description of the use of veneer in furniture manufacture is here included in the hope that it will help the purchaser in identifying and further appraising such furniture as he may be interested in.

By veneered construction is meant the use of veneer in the manufacture of tops, sides, drawer fronts, and other exposed parts of furniture. By solid construction is meant the use of lumber which is not faced by veneers in exposed parts of furniture. The construction of hidden parts, such as drawer bottoms and dresser and mirror backs, does not enter into these definitions.

A large proportion of the furniture made nowadays is veneered. As compared with solid construction veneered furniture has advantages in some respects and disadvantages in others. In order to make clear the relative merits of the two types it is necessary to describe briefly the methods of making each.

In the manufacture of built-up veneer panels such as are used for drawer bottoms and the sides and backs of dressers, buffets, etc., three, or sometimes five, sheets of veneer are glued together with the grain of each extending in a direction at right angles to that of the adjacent sheet. The resulting product is known as plywood.

The end panel in the dresser illustrated in Figure 4 shows the construction of a ply-wood panel of three plies. For exposed parts the surface plies usually are of thin veneer, one twenty-eighth to one-twentieth of an inch in thickness (the " face " ply usually being of a higher grade than the " back " ply), whereas the intermediate ply or plies are thicker and less expensive. For the backing of mirrors and other construction where appearances are not essential, three plies of some moderate-priced wood, each ply usually one-twentieth to one-tenth of an inch in thickness, are used.

As compared with a single board of the same thickness, ply wood has the advantage that its strength and stiffness are more nearly equal along and across the grain of the face and that its shrinking and swelling are relatively very small. Ply wood also is less liable to be split by nails or to be punctured when struck by a sharp object than is a single-ply board of the same thickness; hence ply-wood construction may be made thinner than solid construction, with consequent economy in material.

In making thicker veneered panels, a core of ¾-inch, ⅞-inch, or other suitable thickness is built up of boards glued edge to edge and dressed to a uniform thickness. This type of construction is shown in the dresser top and drawer front of Figure 4. On each side of this core a thin piece of veneer known as the cross banding, usually one-twentieth of an inch in thickness, is glued, with the grain extending at right angles to that of the core. On top of the cross banding on each side is glued a thin sheet of veneer one-twenty-eighth to one-twentieth of an inch thick with its grain parallel to that of the core. The purpose of the cross banding is to reduce transverse shrinking and swelling of the panel, which in turn reduces the chances of checking and of the opening up of glued joints; to reduce curving of the panel, and to even out any small irregularities in the surfaces of the cores. Sometimes only one thickness of veneer is used for each face of the panel; in such cases the grain almost invariably extends at right angles to that of the core, and the face and back plies act as cross banding. In the dresser illustrated in Figure 4 the top has two sheets of veneer on each side, whereas the drawer fronts have only one sheet to each surface.

For cores in furniture stock chestnut, yellow poplar, and basswood are considered excellent, because they are easily dried, worked, and glued, and do not shrink, swell, or warp excessively with changes in atmospheric humidity. Since the core is entirely covered, lumber with certain defects is not objectionable for this purpose. A grade of chestnut containing wormholes is commonly used in preference to the more expensive clear stock. Knots and decayed areas are objectionable, however, because they do not shrink and swell uniformly with the normal wood and may cause unsightly bumps or depressions to develop in the surface after the furniture has been put in use. Red gum (including its sapwood) and tupelo gum are commonly used for core stock, and many other species are used to a more limited extent.

Since lumber in general, and especially plain-sawed lumber of certain species, such as the gums, has a natural tendency to cup and twist when left in wide pieces, it is customary in core manufacture to rip the boards into strips not over 4 or 5 inches wide. These strips are mixed at random, and in some cases every other one is turned end for end. They are then glued together, as illustrated in the dresser top in Figure 4. Thus the natural tendency of the wood to warp out of shape is to a large extent broken up. For this reason, and because of the cross-banding used, a properly constructed veneered table top usually gives less trouble through warping than a solid top.

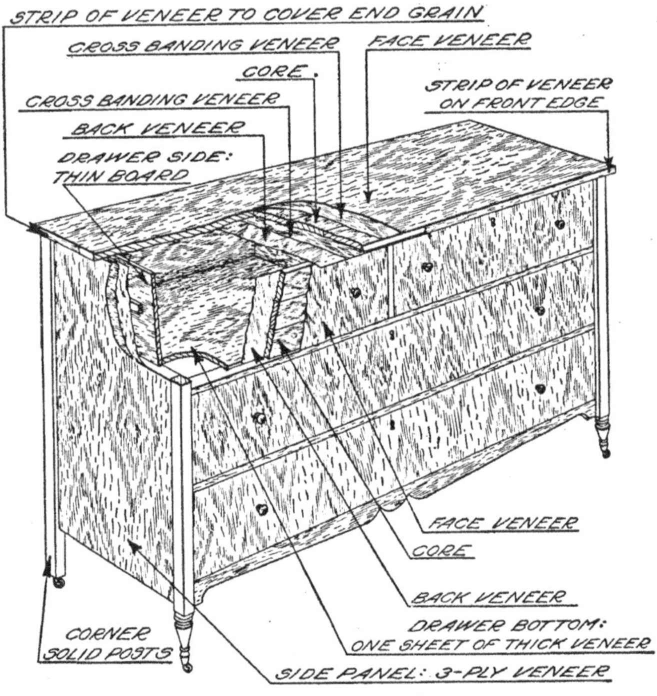

FIG. 4.—Sectional view of medium-grade dresser body, showing use of veneers

A so-called solid top, drawer front, or other part may be composed of one board, but more often consists of several boards glued edge to edge. A strip may be inserted in a groove in the ends of the boards to hold them together, but such a strip usually is not very effective in overcoming the tremendous stresses that occur in lumber as it shrinks and swells with changes in atmospheric humidity. Occasionally, for certain kinds of furniture, such as tables with thick tops and heavy legs, two or more layers are glued together one on top of the other with the grain of all running in the same direction. This is known as laminated construction, but can be classed as solid in contrast to the veneered type.

The advantages of veneered construction as compared with solid construction may be summed up as follows:

1. A ply-wood panel is stronger, in some respects, than a single board of the same thickness.

2. The cores of veneered panels may be made of softer, lighter, and cheaper lumber than can be used for solid construction.

3. Highly figured woods, some of which it would be impractical to use in thick sizes because of their cross grain and resulting irregular shrinkage, can be used for face veneers. This method has the additional advantages of reducing the expense of the use of highly figured woods and of making the supply go farther.

4. Because of the thinness of the finer face veneers, several pieces, cut consecutively, look practically alike and can be matched to produce symmetrical figures impossible to attain in solid construction. (See back cover.)

5. Under ordinary methods of construction, for reasons already explained, veneered panels are less likely to shrink, check, or warp excessively than solid pieces.

6. Curved and irregular surfaces can easily be produced by gluing veneer together in shaped forms which would be difficult if not impossible to produce from solid lumber.

The following are the principal advantages of solid construction:

1. The owner has the satisfaction of knowing that the furniture he possesses is constructed throughout of the kind of wood represented at the surface.

2. The wood can be carved, which is not practicable in veneered construction unless special provision is made for it.

3. In case the surface chips off, the same kind of wood is exposed.

4. The surfaces can be heavily sandpapered or even planed off and refinished—operations which, as a rule, are not practicable with veneered construction.

5. The surface layers can not loosen and peel off, as may occur in veneered panels when they are not properly constructed or are allowed to become wet for any length of time (unless a water-resistant glue is used); although prolonged dampness may likewise have deleterious effects on the glue which is used in the joints of solid furniture.

FEDERAL TRADE COMMISSION RULES FOR FURNITURE NOMENCLATURE

In view of the prevalent use of veneered construction and the common practice of using several kinds of wood for exposed parts of a piece of furniture, the need has been felt for some uniform system of nomenclature that will prevent misunderstanding of terms used in describing the kind of wood of which furniture is made. The Federal Trade Commission early in 1926 approved the following rules and interpretations adopted by leading department stores and other retail dealers in New York City and invited all furniture dealers to subscribe to them.

1. Furniture in which exposed surfaces are of one wood shall be designated by the name of the wood.

2. Furniture in which the exposed surfaces are of more than one kind of wood shall be designated by the names of the principal woods used.

INTERPRETATION OF RULES

1. Exposed surfaces mean those parts of a piece of furniture which are exposed to view when the piece is placed in the generally accepted position for use.

2. The exposed surfaces of all furniture or parts thereof represented as solid shall be of solid wood of the kind or kinds designated. If veneered on the same wood, it may be designated as a wood of that particular kind. If veneered on a different wood it shall be described as veneered.

3. Cabinet woods, used for decorative purposes where the effect is solely to add to the artistic value, shall be named as decorations only.

4. A wood popularly regarded as of lesser value, if its use is essential to construction, need not be named under rule 2, if less than a substantial amount is used on exposed surfaces.

5. A wood popularly regarded as of higher value, shall not be named under rule 2, if an insubstantial amount of that wood is used, except as provided in interpretation 3, above.

6. Designations shall be made in the caption or body of each particular description without qualification elsewhere.

7. The word "finish" to designate color, shall only be used as a description, following the name of the wood used.

8. Where furniture is catalogued, tagged, labeled, advertised, or sold by retailers, it shall be in accordance with these rules and interpretations.

9. Where furniture is catalogued, tagged, labeled, advertised, invoiced, or sold by manufacturers, manufacturers' representatives, jobbers, or wholesalers, it shall be in accordance with these rules and interpretations.

10. The above rules need not apply to antique furniture.

EFFECT OF ATMOSPHERIC CHANGES ON FURNITURE

No matter what kind of wood is used, how well it is selected, seasoned, machined, and put together, or how carefully it is finished with filler, stains, varnish, etc., the fact remains that wood takes on and gives off moisture with variations in atmospheric humidity, and shrinks and swells accordingly.

In temperate climates, in buildings which are heated during winter, the humidity indoors may vary from as low as 10 per cent in winter to as high as 75 per cent in summer. At a temperature of 70° F. and a relative humidity of 10 per cent, wood will eventually attain a moisture content of about 3 per cent of its oven-dry weight; at the same temperature but with 75 per cent relative humidity it will attain a moisture content of 15 per cent. In other words, wood used in buildings heated in winter may be subject to a moisture change with the seasons of 12 per cent of its own weight. Usually the change is not so great, because such extremes of humidity are not common indoors for any length of time; but it is safe to assume that the moisture content of most furniture will vary from about 4 per cent in winter when buildings are heated to 8 per cent or slightly more during other seasons. In very damp climates the maximum moisture absorption would be considerably greater.

Paint, varnish, and enamel do not reduce the amount of moisture absorbed over long periods of time, although such coatings do reduce the rate of absorption and thereby prevent sudden changes in the moisture content of furniture and excessive differences in moisture content between surface and interior. This action is important, since if the moisture is not absorbed fairly uniformly the resulting stresses are apt to cause damage. For example, a table top with several coats of varnish on the upper surface but none on the lower surface will, when exposed to a sudden rise in atmospheric humidity, tend to become dished on account of absorbing moisture and expanding more rapidly on the lower surface than on the upper.

In order to give best results, furniture wood should have a moisture content of about 6 per cent at the time of manufacture into the finished article. Furniture with this initial moisture content will shrink slightly in winter when kept in a heated building, and will swell a little in summer when windows and doors are open; but the shrinking and swelling will not be serious if proper allowance for such changes in dimension is made in manufacturing.

In storing furniture, care should be taken not to expose it to extremely dry air, especially hot dry air; to moist air; or to sudden changes in atmospheric humidities. When wood is properly manufactured and given reasonable care, the gradual humidity changes of the seasons will not, as a rule, affect its quality or usefulness. Careful seasoning methods, the use of quarter-sawed lumber, the selection for core stock of woods which shrink and swell little, the practice of cross-banding in veneered products, and proper finishing procedure will serve in a large degree to eliminate the troubles that might otherwise result from changes of moisture content in the finished article

DESCRIPTION OF WOODS COMMONLY USED FOR EXPOSED PARTS OF FURNITURE

OAK

PRINCIPAL SPECIES OF THE WHITE OAK GROUP

White oak—*Quercus alba* Linn.[5]
Bur oak—*Quercus macrocarpa* Michx.
Swamp white oak—*Quercus bicolor* Willd.
Post oak—*Quercus stellata* Wang.
Swamp chestnut oak—*Quercus prinus* Linn. (formerly *Q. michauxii* Nutt.).
Overcup oak—*Quercus lyrata* Walt.
Chestnut oak—*Quercus montana* Willd.
Chinquapin oak—*Quercus muehlenbergii* Engelm.

PRINCIPAL SPECIES OF THE RED OAK GROUP

Red oak—*Quercus borealis* Michx. and *Q. borealis maxima* (Marsh.) Ashe, (formerly *Q. rubra* of authors, not Linn.).
Black oak—*Quercus velutina* Lam.
Texan oak—*Quercus texana* Buckl.
Southern red oak—*Quercus rubra* Linn. (formerly *Q. digitata* (Marsh.) Sudw.).
Pin oak—*Quercus palustris* Muenchh.
Willow oak—*Quercus phellos* Linn.

---

[5] The scientific names used for native woods in this circular are those which were adopted by the Forest Service late in 1924.

OTHER NAMES

There are 60 species of oak in the United States, but only about 14 of these are of commercial importance as furniture woods. The others are either too small, too defective, or not of sufficient abundance to be used extensively by the furniture industry.

Many of the species are known by different common names in different localities. Thus, swamp chestnut oak is commonly called cow oak or basket oak; chestnut oak, rock oak; chinquapin oak, yellow oak or yellow chestnut oak; black oak, yellow oak; Texan oak, spotted oak or Texas red oak; and Southern red oak, forked-leaf red oak or Spanish oak. Even the scientific names in use are not uniform for each species, though they are far more stable in meaning than the common names. For practical purposes it is not usually necessary to know the exact species of oak lumber, but it is often important to know whether it belongs to the white oak or to the red oak group.

WHERE GROWN

Oak trees and shrubs of one species or another grow over almost the entire wooded area of the United States. The oaks used extensively in the manufacture of furniture, however, are confined to the regions east of the Great Plains. Some oak lumber is manufactured in California and Oregon, but most of the oak used in the West is shipped in from the Eastern States and from Japan.

COLOR AND FIGURE

Heartwood lumber of the white oak group is grayish brown in color, occasionally with a reddish tinge; that of the red oak group usually has a reddish tinge, although sometimes it resembles the white oaks in color. The sapwood, which varies from one to several inches in width, is white in both groups unless discolored by external agencies.

The figure in oak lumber is produced by the annual rings and medullary rays. In plain-sawed lumber the figure consists of straight lines, irregular curves, and parabolas produced by the conspicuous growth rings (pl. 2, *B*).

In quarter-sawed lumber the principal figure is produced by the large medullary rays, which appear as "silver grain" or "flakes" and are more or less continuous across the grain of the wood. (See front cover.) In the white oaks the rays usually average taller than in the red oaks and therefore produce a somewhat more conspicuous figure. When quarter-sawed lumber is not cut perfectly radially the rays appear short in the direction across the grain; when it is cut at a slight angle to the axis of the log, as was done with the specimen used for the front-cover illustration, the full height of the rays does not show.

Only rarely are the fibers in oak contorted so as to form wavy or curly grain, or burls.

STRUCTURE

End surfaces.—The annual rings are very distinct in both the white oaks and the red oaks. Usually, however, those of the red oaks are wider, and also more pronounced because they contain more large pores in the beginning of each year's growth and these pores are more open.

In the heartwood of nearly all white oaks the larger pores are filled with tyloses,[6] whereas in the red oak tyloses as a rule are absent or are only sparsely developed. A more reliable distinction between the woods of the two groups, but one requiring a hand lens for observation, is that in the red oaks the small pores in the dense outer half of each annual ring are individually distinct, and not so numerous but that they may readily be counted, whereas in the white oaks these pores are smaller, more numerous, and difficult to count. Compare Plate 10, *A*, with Plate 11, *A*, for these points of difference.

Although the pores in the denser, outer portions of the annual rings of both white and red oaks are too small to be seen without a magnifying glass, their characteristic arrangement in V-shaped groups or lines extending radially across the rings is apparent to the naked eye.

The oaks have two kinds of rays, relatively large and conspicuous ones and very small ones. Both show in Plates 10, *A*, and 11, *A*, but in unmagnified transverse surfaces only the larger ones appear as visible lines. The size and distribution of the large rays make it easy to distinguish the oaks from all other woods.

Longitudinal surfaces.—On lumber sawed lengthwise the annual rings are made conspicuous, especially in the red oaks, by zones of large pores alternating with zones of denser, less porous wood. The large pores are plainly visible as fine grooves, although in the white oaks the tyloses fill up the grooves more or less.

On plain-sawed surfaces the ends of the large medullary rays can be seen as darker or sometimes lighter colored lines from a fraction of an inch to several inches in height along the grain, as illustrated in the Plates 10, *B*, and 11, *B*.

On quarter-sawed surfaces, if perfectly radial, the large rays appear as patches or "flakes" measuring from a small fraction of an inch to 3 or 4 inches along the grain and the full width of the board across the grain. The small rays are also visible on quarter-sawed surfaces, but they are very inconspicuous (pls. 10, *C* and 11, *C*).

USES IN FURNITURE

Oak is one of the woods most commonly used in furniture both in the form of lumber and as veneer. Its abundance, hardness, strength, and good appearance, as well as its adaptability to various kinds of finishes, qualify it well for all grades of furniture. Although walnut and mahogany may be more in vogue at times, oak is never entirely out of style. The white oaks are preferred for woodwork which is to be given a "natural" finish, because they are usually free from the reddish tinge common in the other group. For darker pieces and parts, however, little discrimination is made between the two.

---

[6] For definition of tyloses, see p. 6.

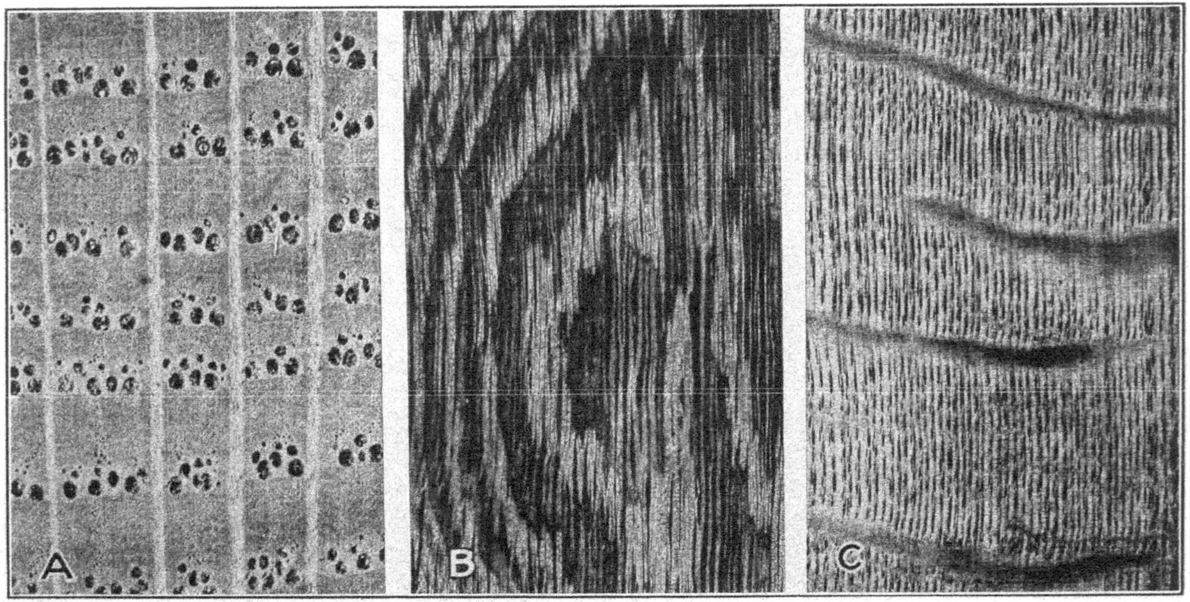

WHITE OAK

A. End surface. (Magnified 7½ diameters.) *B.* Plain-sawed surface. (Natural size.) *C.* Quarter-sawed surface. (Natural size)

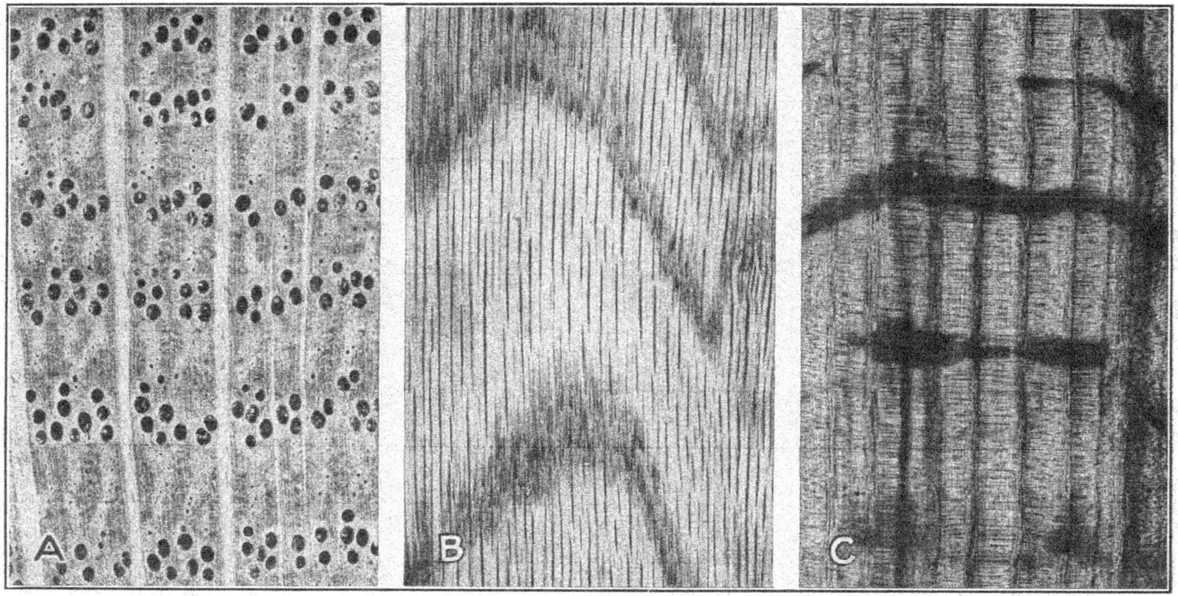

RED OAK

A. End surface. (Magnified 7½ diameters.) *B.* Plain-sawed surface. (Natural size.) *C.* Quarter-sawed surface. (Natural size)

101140°—26——3

CHESTNUT

Castanea dentata (Marsh.) Borkh.

OTHER NAMES

Chestnut is not commonly known by any other name.

WHERE GROWN

Its range is from southern Maine to northwestern Vermont (Winooski River), southern Ontario, and southern shores of Lake Ontario to southeastern Michigan; southward to Delaware and southeastern Indiana, and on the Allegheny Mountains to central Kentucky and Tennessee, central Alabama, and Mississippi.

COLOR AND FIGURE

The heartwood is grayish brown and the thin sapwood almost white. Plain-sawed chestnut has a figure much like that of plain-sawed oak, even more conspicuously outlined by the broad zones of large pores at the beginning of each annual ring. Quarter-sawed chestnut has no especially distinctive figure.

STRUCTURE

End surfaces.—The annual rings are made very conspicuous by several rows of large pores at the beginning of each year's growth. In the outer, denser portion of each annual ring the pores are so small that they can not be seen individually without a lens; but it can be seen that they are arranged in irregular V-shaped or branched radial groups. This arrangement, which is exhibited fairly well in Plate 12, *A*, helps in distinguishing chestnut from ash, with which it is easily confused.

The rays are all very fine, being barely distinct with a hand lens.

Longitudinal surfaces.—On both plain and quarter-sawed surfaces the annual rings are clearly defined. The larger pores are plainly visible as grooves. The rays, however, are not distinct on plain-sawed surfaces and are barely distinct on truly quarter-sawed surfaces (pl. 12, *B* and *C*).

USES IN FURNITURE

Chestnut is used mainly for the cores of table and dresser tops, drawer fronts, and other veneered panels. Figure 4 illustrates the use of cores. For this purpose chestnut is one of the best woods commercially available. Its softness, lightness, ease of drying, and ability to hold glue facilitate manufacturing. Its comparative freedom from warping and the fact that it shrinks and swells very little either during manufacture or in service, give it a stability found in few other species commonly used for cores. When it is used with oak veneer, exposed edges of the core need not be covered, since chestnut is very much like oak in appearance.

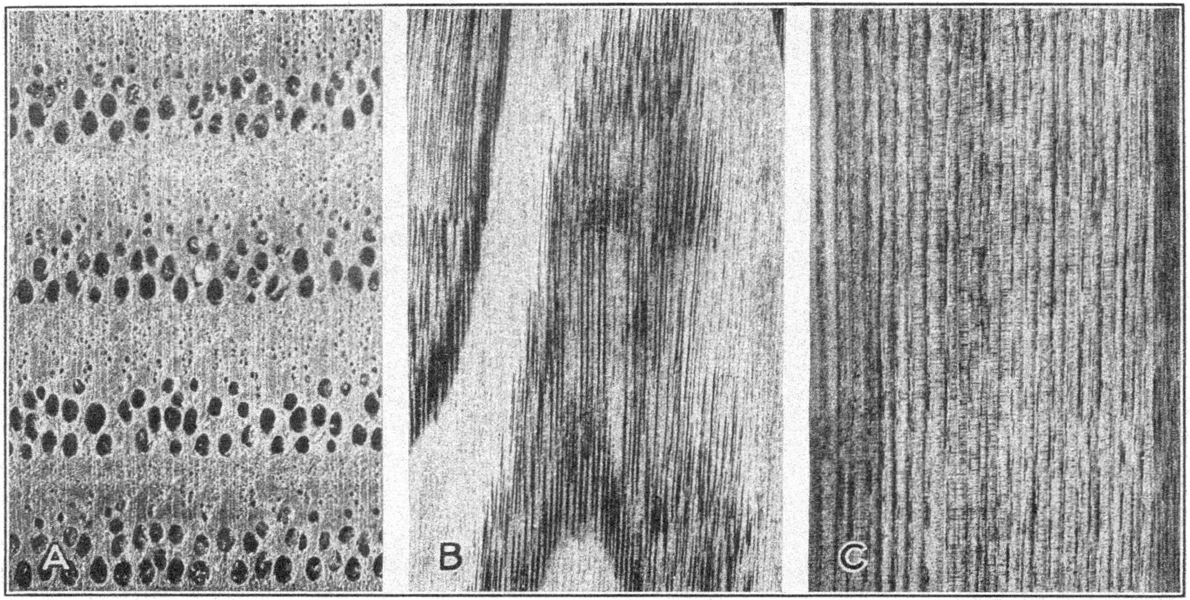

CHESTNUT

A. End surface. (Magnified 7½ diameters.) *B.* Plain-sawed surface. (Natural size.) *C.* Quarter-sawed surface. (Natural size)

A considerable quantity of chestnut timber is infested with wood-boring insects, a condition responsible for a separate grade of lumber known as "sound, wormy chestnut." A limited number of wormholes does not make wood objectionable as core stock and this grade, because of its lower price, can be used to advantage. In some regions the price of chestnut lumber of any grade is relatively low, because of the necessity for cutting the large and increasing proportion of the stand infected with the blight. The blight, however, in no wise affects the quality of the wood if the infected timber is promptly cut.

Chestnut lumber is sometimes used for exposed parts in the cheaper grades of furniture but because of its softness it is undesirable for such parts in the better grades.

ELM

American elm—*Ulmus americana* Linn.
Slippery elm—*Ulmus fulva* Michx.
Rock elm—*Ulmus racemosa* Thomas.

OTHER NAMES

American elm is called also white elm, water elm, swamp elm, and gray elm. Slippery elm is very commonly known as red elm. Rock elm is called also hickory elm, cork elm, and cork-bark elm.

WHERE GROWN

The American elm grows throughout the eastern half of the United States, except the southern tip of Florida. Slippery elm grows in the same region except near the coast of the southern Atlantic and Gulf States. Rock elm grows from Vermont westward through southern Ontario, the Lake States, and eastern Nebraska, and south as far as Memphis, Tenn. Winged elm (*Ulmus alata* Michx.), a species with wood similar to that of rock elm, is of minor commercial importance in the States bordering on the lower Mississippi River and in eastern Texas.

COLOR AND FIGURE

The heartwood of the American and rock elms is light grayish brown, often tinged with red; that of slippery elm is dark reddish brown or chocolate brown. The sapwood of all is white. In slippery elm the sapwood is rarely over one-half inch wide; in the other two species it usually is from 1 to several inches in width. Plain-sawed elm has a conspicuous growth-ring figure and a characteristic delicate wavy figure within each annual ring, as illustrated in Plate 3, *A*. Quarter-sawed, the wood is without figure of any value.

STRUCTURE

End surfaces.—The annual rings of all species of elm are made distinct by a zone of large pores at the beginning of each year's growth. In the American and rock elms this zone consists of one row of pores, as a rule, whereas in slippery elm it consists of several rows, making the last-named species look more porous than the others. These pores are clearly distinguishable without magnification except in rock elm, in which they are near the limit of vision. The smaller pores are arranged in conspicuous wavy tangential lines, but are individually invisible without a lens. This wavy arrangement of the small pores in a distinctive characteristic of the elms. (Pls. 13, *A*, and 14, *A*.)

The rays can not be seen very distinctly without a lens.

Longitudinal surfaces.—On plain-sawed surfaces of elm the typical growth-ring figure is pronounced. The larger pores are individually visible without magnification. Within each annual layer of growth as exposed lengthwise there is a faint wavy pattern consisting usually of from three to eight slightly darker wavy lines.

On quarter-sawed surfaces the larger pores are visible but the rays are very small and inconspicuous. See *B* and *C* of Plates 13 and 14 for the structural characteristics of elm wood.

USES IN FURNITURE

American elm is used to some extent for exposed parts of high-grade furniture. Slippery elm, because of its relative scarcity, and rock elm, because its hardness makes it difficult to work, are relatively little used in the furniture trade.

Elm, although not considered a cabinet wood of the first rank, has a pleasing appearance when plain-sawed and properly filled and varnished. It bends well, a property which makes it especially desirable for curved parts such as chair backs. Its tendency to warp is one of the chief drawbacks to its more extensive use.

ASH

White ash—*Fraxinus americana* Linn.
Green ash—*Fraxinus pennsylvanica lanceolata* (Borkh.). Sargent.
Black ash—*Fraxinus nigra* Marsh.

OTHER NAMES

The most common alternative name for white ash is cane ash. Green ash trees are known also as swamp ash and water ash, but the lumber is sold as white ash. Black ash is often called brown ash, hoop ash, swamp ash, and water ash. Other species of ash grow in this country, but the three here considered constitute 98 per cent of all the ash cut.

WHERE GROWN

White ash grows throughout almost the entire wooded area of the United States east of the Great Plains (except the Gulf and South Atlantic coasts), and in southern Ontario and Quebec.

Green ash has practically the same geographic distribution as white ash, except that in addition it grows along the coast, follows the tributaries of the Mississippi River westward across the prairies almost to their sources, and extends farther northward in Canada.

Black ash grows from along the northern shores of the Gulf of St. Lawrence southward to Delaware, southern Ohio, northeastern Missouri, and northeastern North Dakota.

PLATE XIII

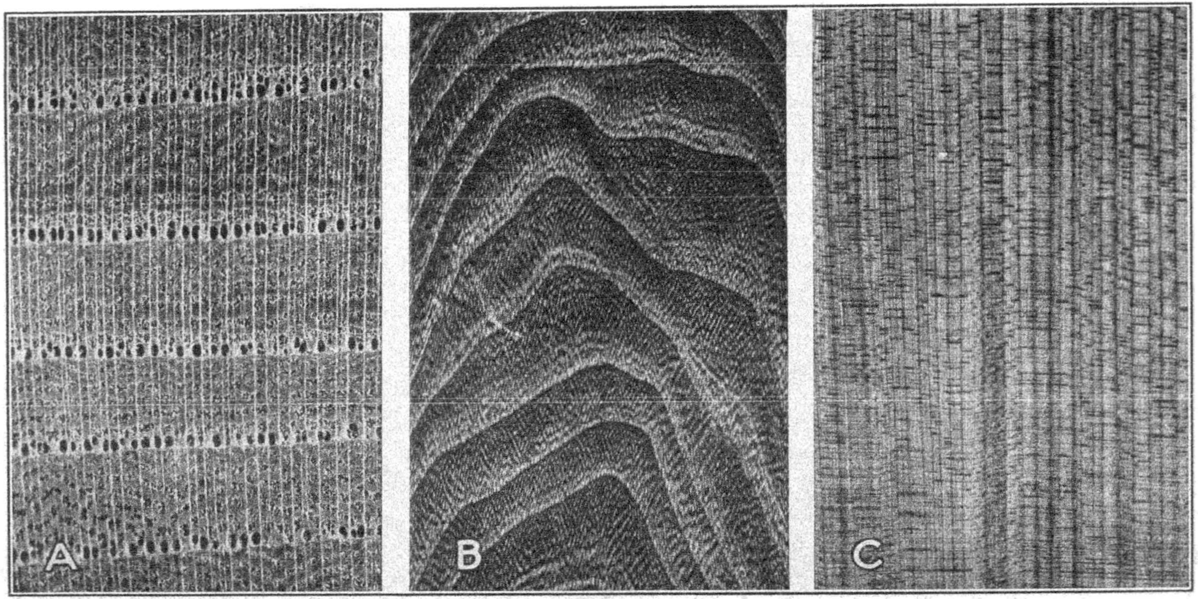

AMERICAN ELM

A. End surface. (Magnified 7½ diameters.) *B.* Plain-sawed surface. (Natural size.) *C.* Quarter-sawed surface. (Natural size)

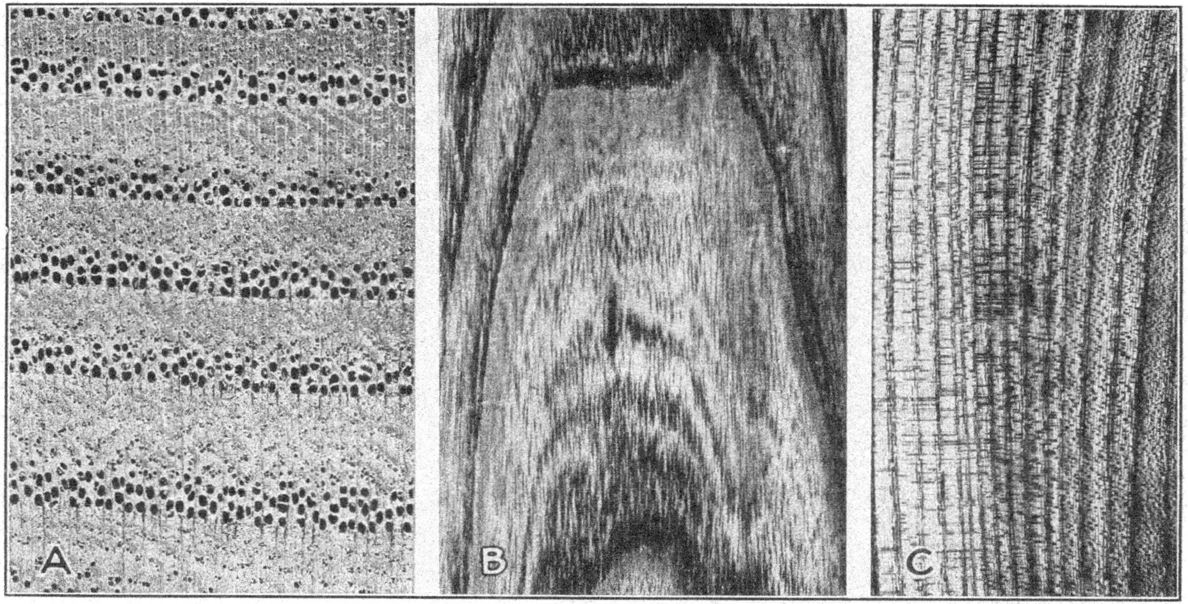

SLIPPERY ELM

A. End surface. (Magnified 7½ diameters.) *B.* Plain-sawed surface. (Natural size.) *C.* Quarter-sawed surface. (Natural size)

COLOR AND FIGURE

The heartwood of white ash and of green ash is light grayish brown in color, sometimes with a reddish tinge. In black ash the heartwood is somewhat darker, hence its name. The sapwood of all three species is white. In the white and green ashes it is several inches wide as a rule, in black ash rarely over 1 inch.

Plain-sawed lumber or rotary-cut veneer exhibits a pronounced figure consisting of ellipses and parabolas formed by the annual rings (pl. 3, *B*). Quarter-sawed lumber has no figure to speak of. Occasionally burls are cut from the trunks and sliced into veneers.

STRUCTURE

End surfaces.—The annual rings are very conspicuously defined by several rows of large pores at the beginning of each year's growth. In the denser outer half of the annual rings the pores are so small that they are not visible without a lens. In this portion of the annual ring in white ash and green ash a few delicate wavy tangential lines may be seen, whereas in black ash these lines usually are lacking. Note the cross section of white ash illustrated in Plate 15, *A*.

The rays are so fine that they are not distinguishable without a lens.

Longitudinal surface.—The annual rings, which are very pronounced, produce a bold growth-ring figure when a dark filler is used. On plain-sawed surfaces an obscure wavy pattern is sometimes noticeable between the zones of large pores, but this pattern is not so pronounced as the otherwise somewhat similar one in elm (pl. 15, *B*).

The rays are not distinctly visible on plain-sawed surfaces and are very inconspicuous on quarter-sawed surfaces (pl. 15, *C*).

USES IN FURNITURE

Ash lumber is not used a great deal for household furniture, being only occasionally worked into tables, dressers, wardrobes, and other pieces of furniture of medium grade. It is used extensively, however, for refrigerators. Ash has properties which are particularly suitable for furniture. It possesses sufficient strength and hardness (black ash ranks lower than white in these respects), but is not too hard to work satisfactorily. It holds its shape well, and it exhibits a good figure when plain-sawed.

PECAN (HICKORY)

Pecan—*Hicoria pecan* (Marsh.) Britton.
Water hickory—*Hicoria aquatica* (Michx. f.) Britton.

OTHER NAMES

The hickory genus is divided into two groups: The true hickories, of which shagbark, big-leaf shagbark, mockernut, and pignut are important species; and the pecan hickories, of which pecan, water hickory, nutmeg hickory, and bitternut are the principal species.

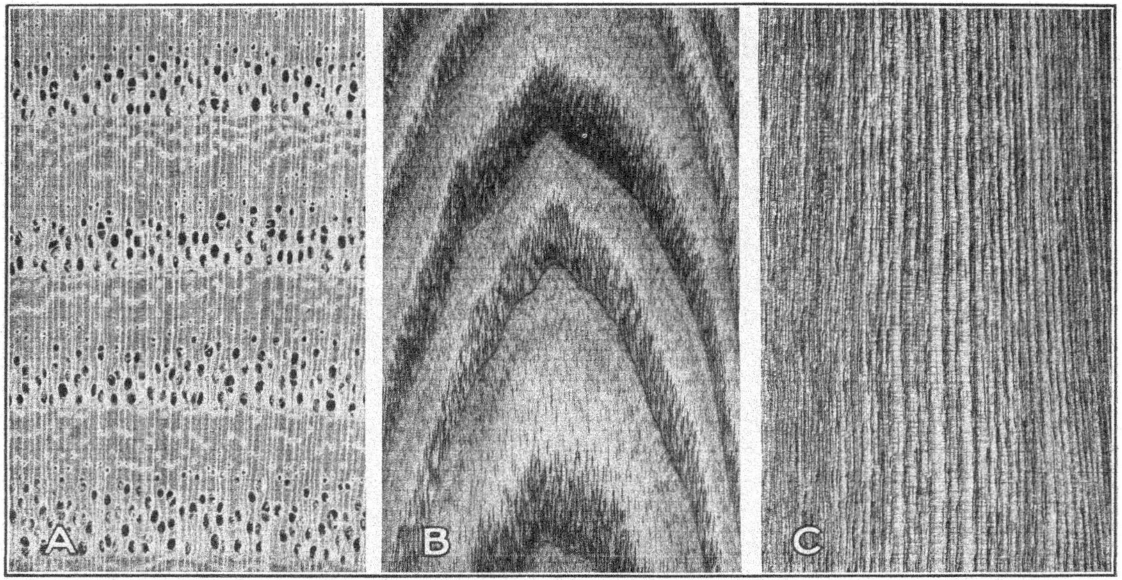

WHITE ASH

A. End surface. (Magnified 7½ diameters.) *B.* Plain-sawed surface. (Natural size.) *C.* Quarter-sawed surface. (Natural size)

The first two species of the pecan group are more important from the furniture-manufacturing standpoint than the others.

Although pecan is a member of the hickory genus, the word "hickory" is rarely used in connection with it. The species is usually referred to simply as pecan. It is sometimes called sweet pecan, and in contradistinction the water hickory is often called bitter pecan.

WHERE GROWN

Pecan grows naturally in the States bordering on the Mississippi River from northern Illinois south to central Louisiana. It extends eastward as far as southern Indiana and western Alabama, and westward as far as eastern Nebraska and central Texas. Its natural range has been considerably extended by cultivation.

Water hickory is most abundant in the lower Mississippi River Valley but grows also along the coast from Virginia to Texas, except in southern Florida.

COLOR AND FIGURE

The heartwood of both species is reddish brown in color, often with darker streaks. The sapwood, which is from 1 to several inches wide, is white. The wood is not highly figured; but it can be so stained and finished as to resemble walnut, to which it is related botanically.

STRUCTURE

End surfaces.—The pores are plainly visible to the naked eye in the inner half of each growth ring. They decrease in size more or less gradually toward the outer margin of each annual ring, and are nowhere crowded.

In water hickory the pores decrease so gradually in size from the inner to the outer limits of each annual ring as to make the wood appear almost diffuse-porous.[7]

The medullary rays are not distinctly visible without a magnifying glass.

With a lens, even one of low power, and occasionally without a lens, 5 to 20 fine, light-colored lines concentric with the rings can be seen within each annual ring. Although fine lines concentric with the annual rings are found also in walnut, oak, and a few other kinds of wood, they are more obscure and not so characteristic as in the hickory genus.

Longitudinal surfaces.—The pores are visible as distinct grooves or lines. The medullary rays on perfectly quarter-sawed surfaces are distinct but not conspicuous; on plain-sawed surfaces they are invisible.

USES IN FURNITURE

Thus far pecan and water hickory have not been used extensively in the manufacture of furniture, but indications are that they will come into greater demand as other woods become scarcer or higher priced. From the manufacturing standpoint, hardness of the wood is the chief drawback to the more extended use of these species, al-

---

[7] For distinction between diffuse-porous and ring-porous woods, see p. 6.

though they are slightly softer than the true hickories. Also, they exhibit a slight tendency to warp.

BIRCH

Yellow birch—*Betula lutea* Michx. f.
Sweet birch—*Betula lenta* Linn.

OTHER NAMES

Yellow birch is known also as gray birch, silver birch, and swamp birch. Sweet birch is known also as cherry birch, black birch, and mahogany birch. The heartwood of both species is usually sold as "red birch," and the sapwood as "white birch" or "yellow birch."

Other species of birch are rarely used for furniture.

WHERE GROWN

Yellow birch grows within an area that extends from Newfoundland westward to northern Minnesota and southward along the Appalachians as far as Asheville, N. C. It is most abundant and reaches its largest size in New England, New York, and northern Michigan and Wisconsin.

Sweet birch occurs from Newfoundland westward to eastern Iowa and southward to northern Florida. The stands of commercial importance are principally in the East, from New York State southward along the Appalachian Mountains.

COLOR AND FIGURE.

The heartwood is from light to dark reddish brown, that of the sweet birch being somewhat darker than that of the yellow. The sapwood, which usually is wide, is practically white.

In quarter-sawed birch the figure, consisting of stripes produced by the annual rings of growth, is in itself monotonous; but the wavy grain which is occasionally present produces a very pleasing natural effect. Plain-sawed birch and rotary-cut birch veneer exhibit a mild pattern formed by the annual rings. In some logs the annual rings follow numerous small depressions and elevations, giving the lumber an undulating figure usually referred to as "curly birch."

STRUCTURE

End surfaces.—The annual rings, though not conspicuous, are clearly defined by fine lines and occasionally by a slight difference in the size of the pores at the end of one year's growth and the beginning of the next.

The pores are almost or entirely invisible without a lens. They are fairly uniform in size, except as already stated, and are evenly distributed, as may be seen from Plate 16, *A*.

The rays are very fine and are not distinctly visible without a lens.

Longitudinal surfaces.—The layers of annual growth are usually distinct but are not so conspicuous as in oak, chestnut, or similar woods with definite zones of large pores.

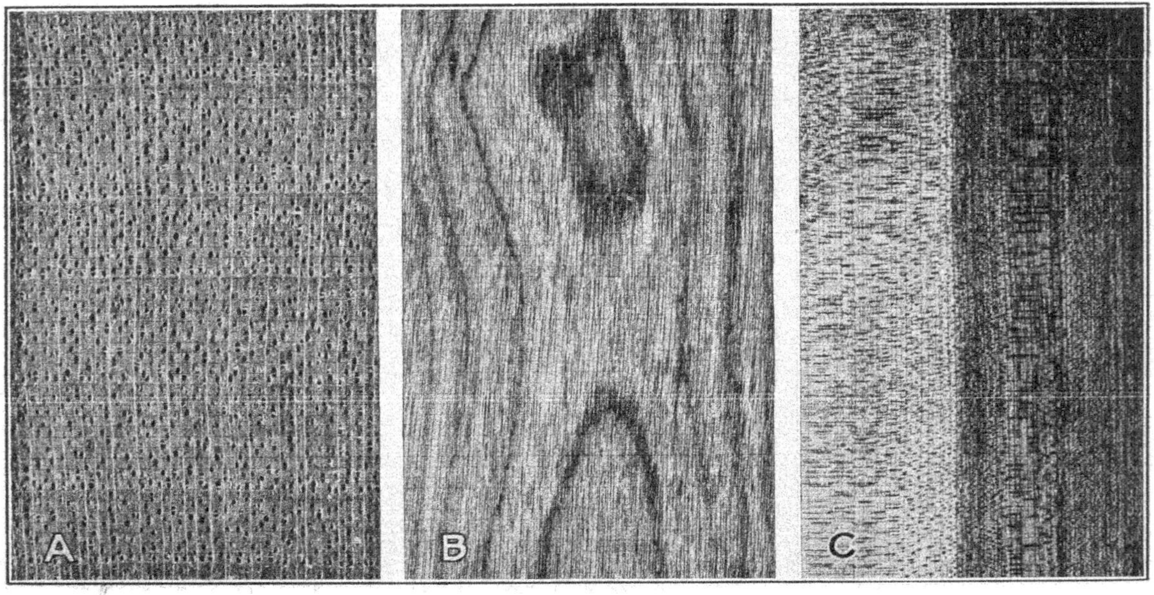

YELLOW BIRCH

A. End surface. (Magnified 7½ diameters.) *B.* Plain-sawed surface. (Natural size.) *C.* Quarter-sawed surface. (Natural size)

The pores can be seen as very fine grooves or lines on smoothly planed surfaces and even on varnished surfaces, especially if a dark filler has been applied (pl. 16, *B*).

The rays are not visible on plain-sawed surfaces but appear as very small reddish-brown " flakes " on smooth quarter-sawed surfaces, as shown in Plate 16, *C*.

There is not sufficient difference between the structure of yellow birch and that of sweet birch to make it possible always to tell these two apart.

USES IN FURNITURE

Both yellow and sweet birch are used extensively in the manufacture of furniture. Of the two the yellow birch is used in larger quantities, but only because of its greater abundance. Birch is used to some extent in a " natural " finish but often is stained reddish brown to imitate mahogany or chocolate brown to imitate walnut. Manufacturers also esteem it highly for furniture which is to have an enamel finish. A large percentage of the birch that goes into furniture is applied in the form of veneers.

The physical properties of birch make it particularly suitable for furniture manufacture. It is sufficiently strong and hard, but not unnecessarily hard, and it holds its shape well. Although it shrinks more than lighter woods, this is not a serious handicap if the lumber is properly dried.

ROSEWOOD[8]

Dalbergia nigra Fr. Allemão

OTHER NAMES

The name " rosewood," as used commercially, is applied to a large number of species. As in the case of mahogany, some of the woods so called are not even remotely related botanically to the original. True Brazilian rosewood is botanically known as *Dalbergia nigra* Fr. All. It belongs to the legume family. Locally it is called jacarandá.

Other species of *Dalbergia* also are classed as rosewood by the trade, as for example Honduras rosewood (*Dalbergia* sp., specific name as yet undetermined), Indian rosewood (*Dalbergia latifolia* Roxb.), and Madagascar rosewood (probably *Dalbergia baroni* Baker). Cocobolo wood, which also belongs to the genus *Dalbergia*, is occasionally exported from Central America under the name of rosewood,[9] but because of the small size of the logs is not used in furniture.

Several species of the closely related genus *Machærium* are exported from Brazil as rosewood, to which they have a very close resemblance. In this circular no attempt is made to distinguish the woods of the various South and Central American species of *Dalbergia* and *Machærium*, since there is more or less uncertainty at present as to the botanical classification of the trees themselves.

Other names for tropical American rosewood are Brazilian rosewood, Bahia rosewood, Rio rosewood, palisander wood, and cabiuna.

---

[8] For part of the information on rosewood the author is indebted to Record and Mell's " Timbers of Tropical America," 1924.

[9] RECORD, S. J., and GARRATT, G. A., 1923. COCOBOLO. Yale School of Forestry. Bul. 8.

The name "rosewood" originated from the roselike odor of true rosewood. This odor is comparatively faint in the seasoned wood, but very pronounced in the green lumber.

WHERE GROWN

True rosewood grows throughout the greater portion of the hardwood forests of eastern Brazil.

COLOR AND FIGURE

The heartwood of true rosewood varies greatly in color, ranging from a plain to a highly variegated brown. The kind that is considered most valuable for export has a dark reddish brown or purplish brown ground color interspersed with irregular black or brownish black streaks. The figure of the wood is due almost entirely to these dark streaks, which as a rule do not follow the grain but describe irregular curves—often of fanciful design.

The sapwood, which is white, is usually hewn off the logs before they are exported.

STRUCTURE

End surface.—Growth rings, which perhaps do not correspond strictly to annual periods, are present but not conspicuous.

The pores are visible to the unaided eye. Many of those in the heartwood are filled with dark-colored gum and are visible only as a result of the reflection of light from the gum. The pores are for the most part scattered singly and are not crowded.

The rays are very fine and not distinctly visible without a lens. With a lens very numerous fine, light-colored lines can be seen extending tangentially from ray to ray and forming with the rays a very delicate network (pl. 17, *A*).

Longitudinal surfaces.—The pores can be seen distinctly on smoothly cut radial and tangential surfaces, appearing as fine grooves mostly filled with glistening gum.

Plain-sawed surfaces under certain light conditions show very fine "ripple marks" produced by the storied arrangement of the medullary rays and other elements.[10] Often these ripple marks are invisible without a lens. On quarter-sawed surfaces the rays are very small and obscure.

USES IN FURNITURE

The use of rosewood for furniture has been gradually decreasing for the last several decades, largely on account of changes in styles, although its high price and the fact that it is comparatively hard to work may be contributory to its decline. Its principal household use at present is in piano cases, musical instruments, handles, brush backs, and novelties.

MAHOGANY

Five species of mahogany have been identified by botanists.[11] Of these the two most important are *Swietenia mahagoni* Jacq. and *Swietenia macrophylla* King.

---

[10] See p. 7.

[11] For additional information in this connection see the author's publication, "The Identification of True Mahogany, Certain So-called Mahoganies, and Some Common Substitutes," published as U. S. Dept. Agr. Bul. No. 1050.

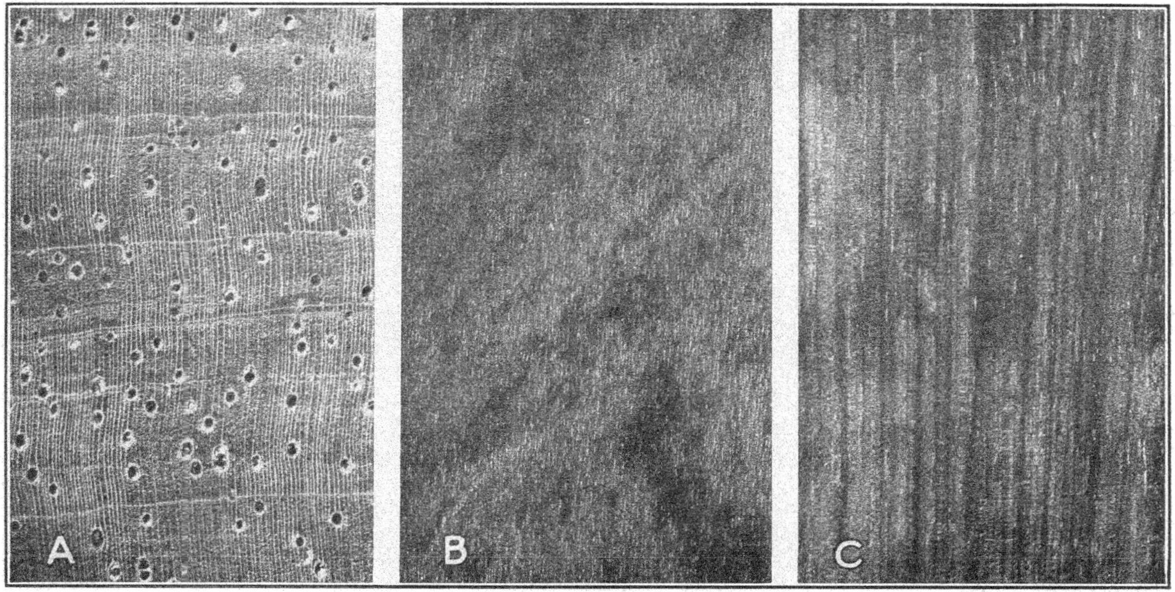

ROSEWOOD

A. End surface. (Magnified 7½ diameters.) *B.* Plain-sawed surface. (Natural size.) *C.* Quarter-sawed surface. (Natural size)

OTHER NAMES

Mahogany is rarely known by any other name. The light-colored soft wood from the east coast of Mexico and of Central America is sometimes called "baywood," and the dark-colored harder wood from the West Indies was called Spanish mahogany in the early days. The latter term is now practically obsolete. Sometimes the wood is classified as Cuban mahogany, Honduras mahogany, Mexican mahogany, etc., according to the country it comes from; but if the place of origin is not known, it can not be determined from an examination of the wood.

WHERE GROWN

Swietenia mahagoni grows in the West Indies, Bermuda, and the keys of southern Florida. *Swietenia macrophylla* grows along the eastern coast of tropical America from the State of Tabasco, Mexico, to Honduras and possibly farther south. Of the other three species of *Swietenia* known, two grow on the west coast of Mexico and of Central America and one is a native of Venezuela.

COLOR AND FIGURE

The heartwood of mahogany varies in color from a pale to a deep reddish brown, becoming darker on exposure to light. The narrow sapwood is from white to light brown in color.

The predominating figure in mahogany is due to differences in the reflection of light from adjacent portions of the surface. Since mahogany nearly always has interlocked grain, quarter-sawed surfaces usually have a ribbon, or stripe, figure. Occasionally mottled, fiddle-back, raindrop, roe, and curly figures are present in the quarter-sawed lumber and enhance its value. Lumber and veneer cut from crotches in the tree trunk also have a distinctive figure and a superior value. Plain-sawed mahogany has a figure of soft outlines and low contrast as a rule, although certain types of figure, when present, show to best advantage in such lumber—blister figure and burls, for example. For various patterns produced by distortion of the grain see *A* and *B* of Plates 6, 7, and 8. A plain-sawed surface is shown in Plate 18, *B*.

STRUCTURE

End surfaces.—Growth layers varying from one-sixteenth to one-half of an inch or more in width are seen on transverse surfaces of mahogany, bounded by narrow, light-colored concentric lines. The pores, plainly visible, are fairly uniform in size and evenly distributed. Many contain dark reddish-brown gum. Occasionally, especially in the heavier grades of the wood, the pores contain white deposits. The rays, though very fine, are distinctly visible without a lens. These features are shown in Plate 18, *A*.

Longitudinal surfaces.—The light-colored layers defining the growth rings are distinct, forming approximately straight lines on quarter-sawed surfaces and curved lines or ovals on plain-sawed surfaces. In Plate 18, *B*, these layers appear dark, because of the way in which the light was reflected.

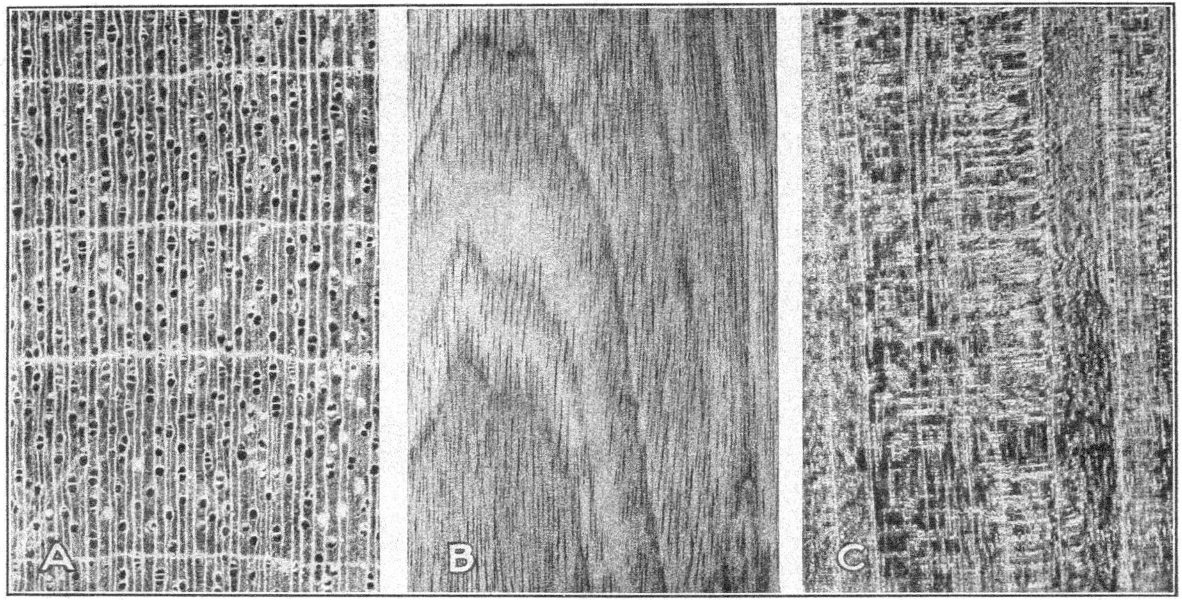

TRUE MAHOGANY

A. End surface. (Magnified 7½ diameters.) *B.* Plain-sawed surface. (Natural size.) *C.* Quarter-sawed surface. (Natural size)

101140°—26——4

The pores are visible as small grooves, often partly filled with dark gum.

The rays are very distinct on quarter-sawed surfaces, where they may appear lighter or darker than the wood fibers according to how the light strikes them. On plain-sawed surfaces they often are in rows running at right angles to the fibers, producing very fine "ripple marks" (pl. 18, *B*). The distinctness and abundance of such "ripple marks" help to distinguish the true mahogany from khaya and from most other woods sometimes called mahogany, but their absence does not disprove genuineness.

USES IN FURNITURE

True mahogany is one of the principal woods used in high-grade furniture of all kinds. It is used most often in the form of veneer, although some of the very best furniture and many of the real antique pieces are of the solid lumber. In less expensive sets red gum and birch are commonly used in combination with mahogany. In such cases the less conspicuous parts, such as the corner posts and rails of dressers, chiffoniers, beds, etc., and the legs of tables and chairs, are made of solid gum or birch, and the broader surfaces are veneered with mahogany.

The principal attributes of mahogany which have given it its prominence as a furniture and cabinet wood are its variety of pleasing figures, its slight shrinkage, its stability, and its ease of working.

KHAYA

Khaya senegalensis A. Juss., and other species of *khaya*

OTHER NAMES

These species have also been called African mahogany, Senegal mahogany, Gambia mahogany, Gaboon mahogany, Benin mahogany [12]—the last four names indicating regions from which the wood is obtained.

WHERE GROWN

Their range is the west coast of Africa and inland along a belt from 15° north to 20° south of the Equator. They also occur occasionally in Uganda and Mozambique on the east coast, and in Madagascar.

COLOR AND FIGURE

Khaya belongs to the mahogany family and is very much like the true, or American, mahogany in color and figure except that the color usually is not so light as in the so-called baywood. As in true mahogany, the color darkens on exposure.

The stripe figure, a characteristic of true mahogany, is fully as pronounced in khaya. Occasionally other types of figure, such as mottle, raindrop, and blister pattern, are also present.

---

[12] The name mahogany in the opinion of the Forest Service should be confined to the genus *Swietenia*.

STRUCTURE

End surfaces.—A distinguishing characteristic of khaya is that the seasonal growth rings are not defined by zones of larger pores or thin lines of lighter or darker colored tissue. Zones of slightly denser wood may alternate with more porous zones but are very irregular in occurrence.

The pores of khaya average slightly larger than those of true mahogany and give the wood a slightly coarser texture; but the difference is not enough to afford a reliable distinction between the two (pl. 19). The pores are fairly uniformly distributed singly or by twos, with occasionally three or four in a radial row. The rays are barely visible without a lens.

Longitudinal surfaces.—In lengthwise section the pores are plainly visible as fine grooves, many of which contain dark-colored gum. On plain-sawed surfaces ripple marks, due to storied arrangement of the rays, are usually lacking and are never present over large areas. On quarter-sawed surfaces the rays appear small and of about the same color as the surrounding wood, but differences in reflection may make them appear darker or lighter.

USES IN FURNITURE

Khaya is used for practically the same purposes as true mahogany from the American Tropics. There is little difference in the price and quality of corresponding grades of the two woods.

TANGUILE, RED LAUAAN, AND ALMON

Tanguile—*Shorea polysperma* Merr.
Red lauaan—*Shorea negrosensis* Foxw.
Almon—*Shorea eximia* Scheff.

OTHER NAMES

Tanguile is sometimes called Bataan mahogany or tanguile mahogany, but with these exceptions the woods have been usually sold without distinction as "Philippine mahogany" in this country.

WHERE GROWN

The Philippine Islands are the source of lumber of these species.

COLOR AND FIGURE

In tanguile and red lauaan the heartwood varies in color from pale to dark reddish brown. It usually has a slight purplish tinge, especially in tanguile. Almon is almost always light in color, sometimes even straw-colored. Unstained surfaces of these species become paler on exposure to light. Their sapwood is very pale grayish or reddish brown.

These woods as a rule show pronounced interlocking of grain, which gives a conspicuous ribbon or stripe effect to quarter-sawed surfaces (pl. 5, *A*). Wavy grain is also occasionally found.

PLATE XIX

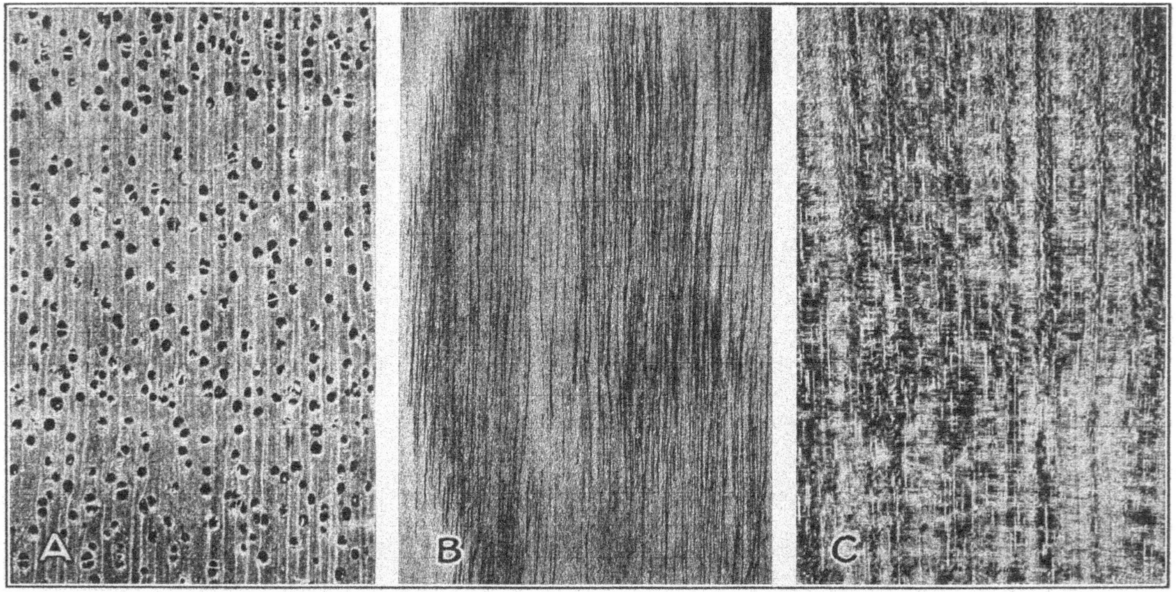

KHAYA

A. End surface. (Magnified 7½ diameters.) *B.* Plain-sawed surface. (Natural size.) *C.* Quarter-sawed surface. (Natural size)

STRUCTURE

End surfaces.—No well-defined annual rings are present. The pores are plainly visible without a lens, evenly distributed, and fairly uniform in size in each species, but slightly smaller on the average in tanguile than in red lauaan and almon. They contain abundant, glistening tyloses. Tangential lines almost white in color (consisting of rows of minute gum ducts) and varying in length from very short to the full width of a board or plank, spaced from one-eighth of an inch to an inch or more apart, are usually visible on smoothly cut end surfaces. The rays are not distinctly visible without a lens (pl. 20, *A*).

Longitudinal surfaces.—In lengthwise section the pores are plainly visible as fine grooves, which on quarter-sawed surfaces may be very short in places owing to interlocking grain.

On quarter-sawed surfaces the rays, though not large, are very conspicuous because of their reddish color. Plate 20, *B* and *C*, show the longitudinal surfaces.

USES IN FURNITURE

These woods, both as lumber and as veneer, have been used to some extent for furniture in this country for several decades. Although they resemble true mahogany considerably, they have an individuality which enables one to distinguish them easily after one has once become familiar with them. This is only to be expected since they belong to a different botanical family.

As furniture woods these species compare favorably in strength with mahogany, but they shrink and swell slightly more with equal changes in moisture content and have a greater tendency to warp when plain-sawed. Differences in color have already been pointed out.

BLACK WALNUT

Juglans nigra Linn.

OTHER NAMES

Walnut; American walnut.

WHERE GROWN

The natural range of black walnut is from southern Ontario to Florida, central Alabama and Mississippi, and westward through southern Michigan, Wisconsin, and Minnesota to Nebraska, Kansas, and Texas (San Antonio River). Its commercial range is restricted to the States bordering on the Ohio and central Mississippi Rivers.

COLOR AND FIGURE

The heartwood varies in color from light to dark chocolate brown, sometimes with a slight lavender tinge. Occasionally dark, irregular streaks are present, running more or less lengthwise of the grain. The narrow sapwood is pale brown; sometimes it is artificially darkened by steaming to make it resemble the heartwood.

The figure of straight-grained quarter-sawed lumber consists of stripes formed by the annual layers of growth or by zones differing

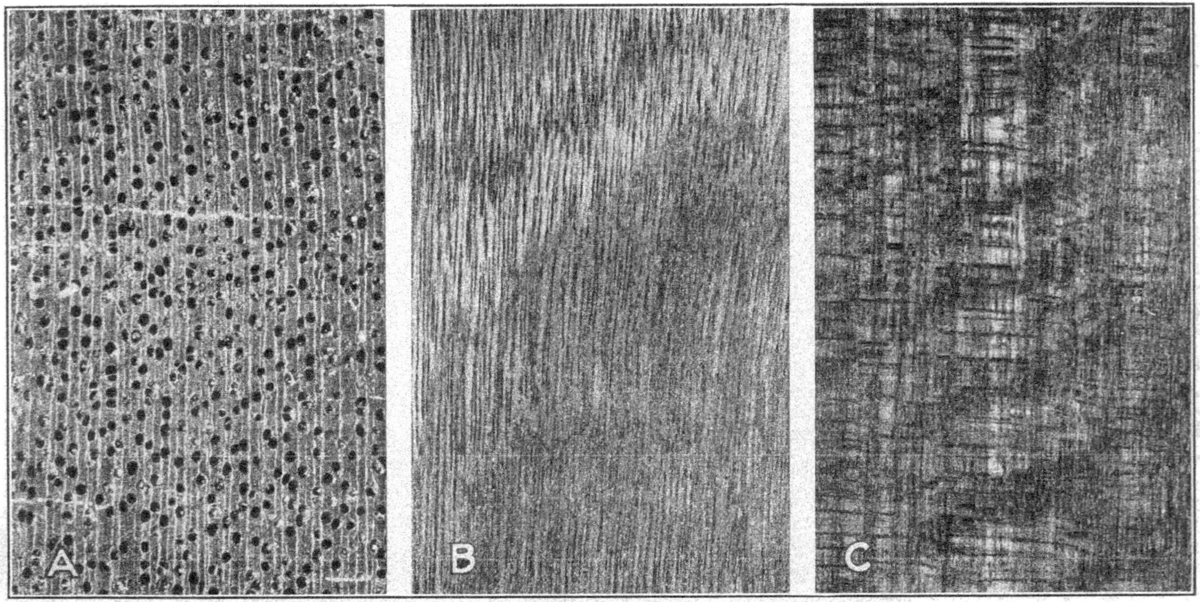

TANGUILE

A. End surface. (Magnified 7½ diameters.) *B*. Plain-sawed surface. (Natural size.) *C*. Quarter-sawed surface. (Natural size)

slightly in shades of color. The ribbon grain so common in mahogany is rarely present, since black walnut seldom develops interlocked grain. Occasionally it shows a wavy figure. In plain-sawed lumber the figure consists of stripes, irregular curves, and parabolas outlining the annual layers of growth, but is mild in comparison with the high contrasts of oak or ash. Veneer cut from stumps and crotches usually shows the irregular figure known as mottle or "butt wood," consisting of short waves and curly contortions of the fibers. (See back cover.)

Burls are highly figured with small circlets and curly grain and for this reason bring a good price. The wood of black walnut is lustrous, a characteristic which adds much to the value of its figure.

STRUCTURE

End surfaces.—The annual rings are distinct, being marked by an abrupt difference in the size of the pores in adjoining portions of adjacent rings.

The pores are comparatively large and are easily visible to the naked eye. They gradually decrease in size from the inner to the outer portion of each annual ring. Most of them contain glistening tyloses.

The rays are so fine and inconspicuous that they are not seen distinctly without a lens. At slight magnification there are discernible not only the rays but also numerous fine tangential lines forming a network with them. Plate 21, *A*, shows these details.

Longitudinal surfaces.—In lengthwise section the annual rings are clearly marked but not conspicuous.

The pores are visible as fine grooves or dark lines on both the plain-sawed and quarter-sawed cuts.

The rays are very fine and inconspicuous, even on quarter-sawed surfaces. See Plate 21, *B* and *C*, for characteristics of longitudinal surfaces.

USES IN FURNITURE

Black walnut has long been a popular wood for furniture. It possesses practically all the properties desirable for that purpose—(1) sufficient hardness and strength for general use but not hardness enough to dull woodworking tools excessively, (2) rich color and luster, (3) a distinctive but not obtrusive figure, (4) comparative freedom from warping, and (5) good gluing qualities. Walnut furniture is manufactured extensively in both solid and veneered construction, the latter particularly where matched or highly figured panels are desired.

Red gum, stained to match, is often used in combination with walnut. It can be distinguished, however, by the absence of visible pores on the surface.

CIRCASSIAN WALNUT

Juglans regia Linn.

OTHER NAMES

Circassian walnut, on account of its wide distribution, enjoys a varied nomenclature. Among the names applied to it are English

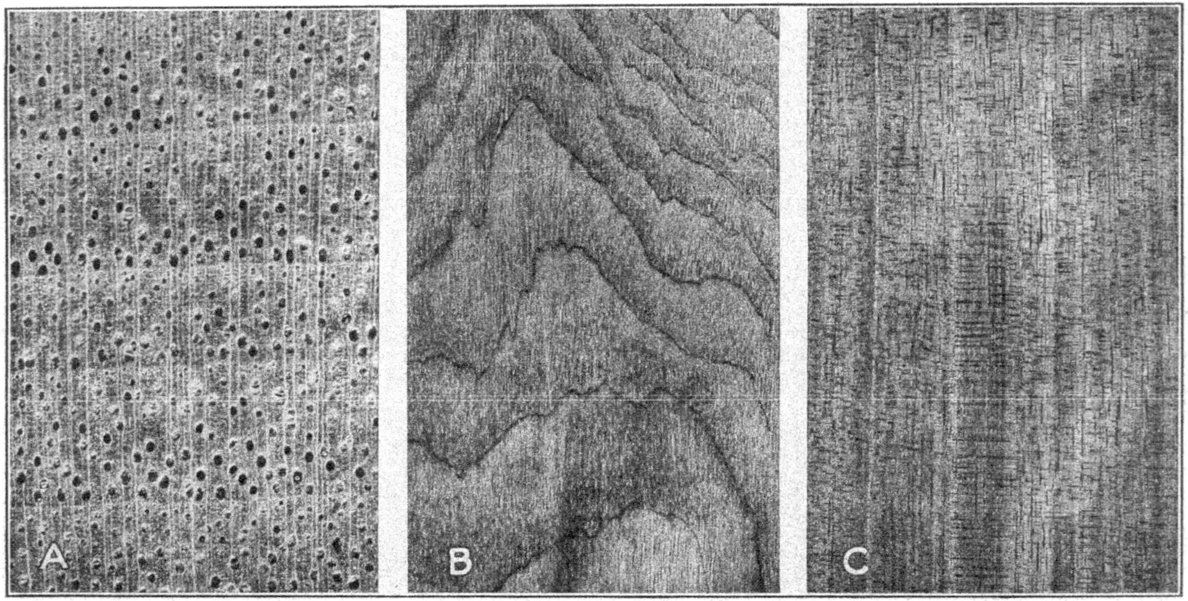

BLACK WALNUT

A. End surface. (Magnified 7½ diameters.) *B.* Plain-sawed surface. (Natural size.) *C.* Quarter-sawed surface. (Natural size)

walnut, royal walnut, Italian walnut, European walnut, French walnut, Persian walnut, Austrian walnut, Turkish walnut, and Russian walnut.

WHERE GROWN

This species grows wild in the Balkans, Asia Minor, the Caucasus Mountains, Persia, and eastward through the Himalayas to Burma and parts of China and Japan. It has been planted extensively in Germany, France, England, India, China, the United States, and other countries, both for its wood and for its nuts (English walnuts). The wood of planted trees, especially those remote from the natural range of the species, is not of such good quality as that of natural forest growth. Up to the time of the World War the principal source of Circassian walnut timber shipped to this country was the region bordering on the Black Sea; in fact it is from Circassia, a Russian district of that region, that the common name is derived.

COLOR AND FIGURE

Fundamentally the heartwood is fawn colored. In addition to this tone, numerous dark-brown or black streaks usually are present. The abundance and irregularity of these streaks determine to a large extent the figure of the wood (pl. 4, *A*). The sapwood is of a pale fawn color or almost white.

STRUCTURE

End surfaces.—Circassian walnut resembles black walnut in structure. The pores are plainly visible without a lens. They decrease in size gradually from the inner to the outer limit of each year's growth, thus defining the annual rings. The rays, unlike those of black walnut, can be seen fairly distinctly without a lens. With a lens numerous fine light-colored tangential lines can be seen on a smoothly cut end surface.

Longitudinal surfaces.—In lengthwise section the pores appear as well-defined grooves. The annual layers of growth are not pronounced, being obscured by the dark streaks previously referred to, the trend of which is independent of annual rings or any other part of the structure (pl. 4, *A*). The rays are small and of about the same color as the rest of the wood. Plate 22 illustrates these structural features of Circassian walnut.

USES IN FURNITURE

At one time Circassian walnut was used to a considerable extent in the manufacture of high-grade furniture in the United States. It was used principally in the form of veneers, although in the highest grade of furniture the corner posts, mirror frames, legs, and other massive or carved parts were solid. In many pieces of furniture it was used in combination with red gum, which, except for its smaller pores, bears a close resemblance to Circassian walnut. During the World War imports of this wood were cut off, and shipments on a commercial scale have not been resumed.

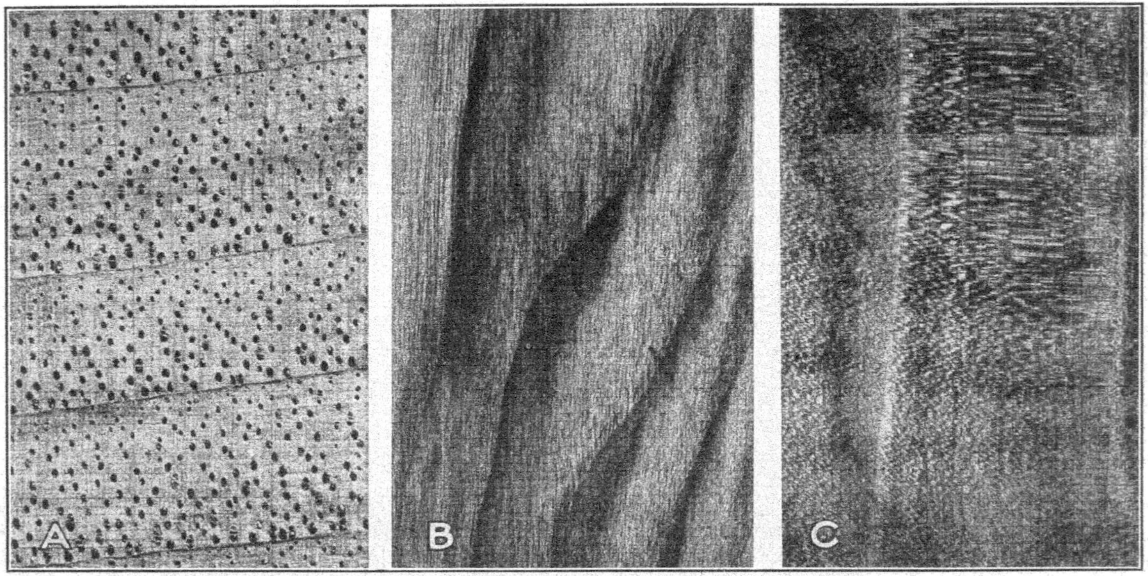

CIRCASSIAN WALNUT

A. End surface. (Magnified 7½ diameters.) *B.* Plain-sawed surface. (Natural size.) *C.* Quarter-sawed surface. (Natural size)

SYCAMORE

Platanus occidentalis Linn.

OTHER NAMES

Plane tree; buttonwood; buttonball tree.

WHERE GROWN

The commercial species of sycamore grows from southern Maine west to eastern Nebraska and Kansas, and south to northern Florida, central Alabama and Mississippi, and eastern Texas.

COLOR AND FIGURE

The sapwood, which usually is narrow, is of a pale reddish-brown tint. The heartwood is deeper reddish brown in color but is not sharply defined from the sapwood.

Sycamore does not present a figure of any value except on quarter-sawed surfaces, where the comparatively large and reddish rays produce a rather striking pattern.

STRUCTURE

End surfaces.—The annual rings, although not conspicuous, are clearly differentiated by light-colored lines which represent the termination of each year's growth.

The individual pores are not visible without a lens. With a lens they appear very numerous and fairly evenly distributed except in the narrow outer portion of the annual ring. There they are less numerous, or at any rate smaller, giving that portion a less porous appearance.

Next to oak, sycamore has the largest rays of any of our native commercial woods. They appear as numerous distinct lines on the cross section, which is illustrated in Plate 23, *A*.

Longitudinal surfaces.—The annual rings are not conspicuous in lengthwise section, although the whitish lines limiting the growth rings are usually visible. The pores are not visible.

On tangential surfaces the rays are distinctly visible as crowded dashes up to one-fourth of an inch in height, but mostly not over three-sixteenths of an inch, darker than the surrounding wood (pl. 23, *B*).

On radial surfaces the rays are very conspicuous on account of their comparatively large size and darker color. Some of them reach one-fourth of an inch in height, but more often they are from one-sixteenth to three-sixteenths of an inch. If a sycamore board is perfectly quarter-sawed many of the rays will extend across the full width of the face, but usually either because the "quartered" lumber is cut at a slight angle with the rays or because the rays curve slightly in and out, they appear as short "flakes" (pl. 23, *C*).

USES IN FURNITURE

Sycamore is not used to any great extent in furniture. It is not very abundant in large sizes and the trees often are defective.

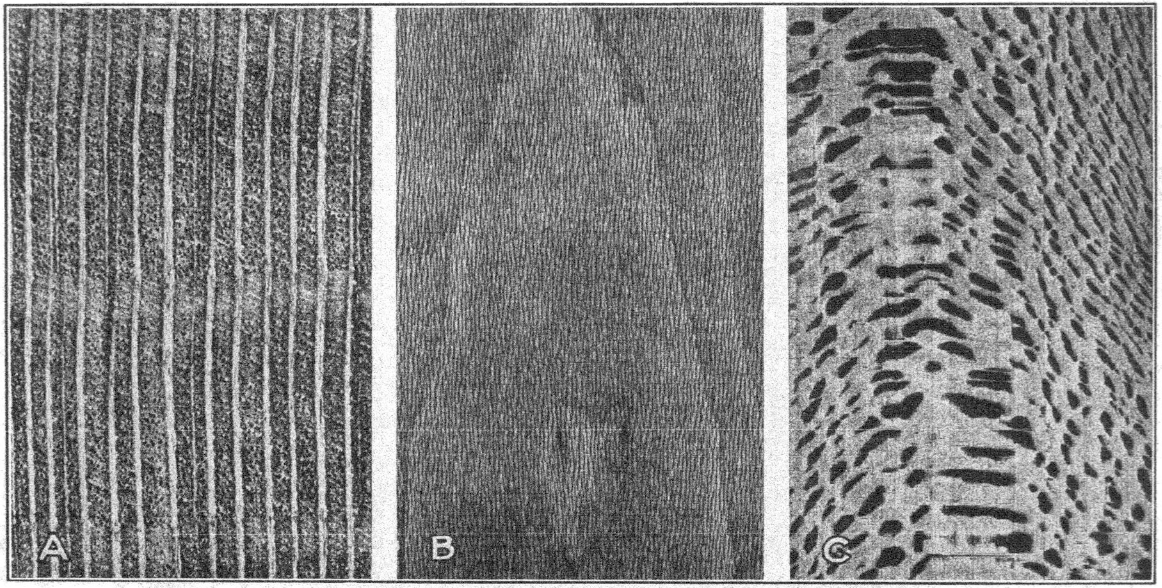

SYCAMORE

A. End surface. (Magnified 7½ diameters.) *B.* Plain-sawed surface. (Natural size.) *C.* Quarter-sawed surface. (Natural size)

Furthermore the lumber has a pronounced tendency to warp. When properly dried and finished, however, it is as well suited for furniture as the wood of the red gum tree, which it approximates in strength, hardness, and shrinkage.

BEECH

Fagus grandifolia Ehrh.

OTHER NAMES

Beech is rarely known by any other name. Trees in which the heartwood is reddish are called " red-heart beech," those with white heartwood " white-heart beech."

WHERE GROWN

The range of the species includes the territory east of a line from the middle of the northern peninsula of Michigan to Galveston, Tex., with the exception of the peninsula of Florida. It also includes southeastern Canada.

COLOR AND FIGURE

The wood of beech is white or slightly reddish. The heartwood is not clearly defined from the sapwood. Beech has no figure to make it ornamental, and it is rarely used for the exterior of furniture other than chairs.

STRUCTURE

End surfaces.—The annual rings are defined by comparatively narrow bands which are slightly darker and less porous.

The pores are not visible without a lens. They are numerous in the inner part of each annual ring, decreasing in size and number toward the outer part.

Some of the rays are very conspicuous, appearing under a lens fully twice as wide as the largest pores, whereas others are very fine. See Plate 24, *A*, for details of the cross section.

Longitudinal surfaces.—On plain-sawed surfaces, the annual rings are inconspicuous. The pores are not visible, but the larger rays appear as short, darker dashes up to one-eighth of an inch in height.

On quarter-sawed surfaces the rays are conspicuous as darker " flakes " from one-sixteenth to one-eighth of an inch in height and of a varying length depending on how nearly radially the lumber is cut. These, however, are not large enough to give the wood a figure of any value.

USES IN FURNITURE

Beech is used in the manufacture of chairs and exterior parts of furniture, which are painted or stained so as to obscure the identity of the wood. It bends easily and is therefore well adapted for curved parts, such as the backs of chairs. Because of its hardness it is also well suited for rocker runners. It is used for drawer sides, drawer runways, frames, and other substantial interior parts.

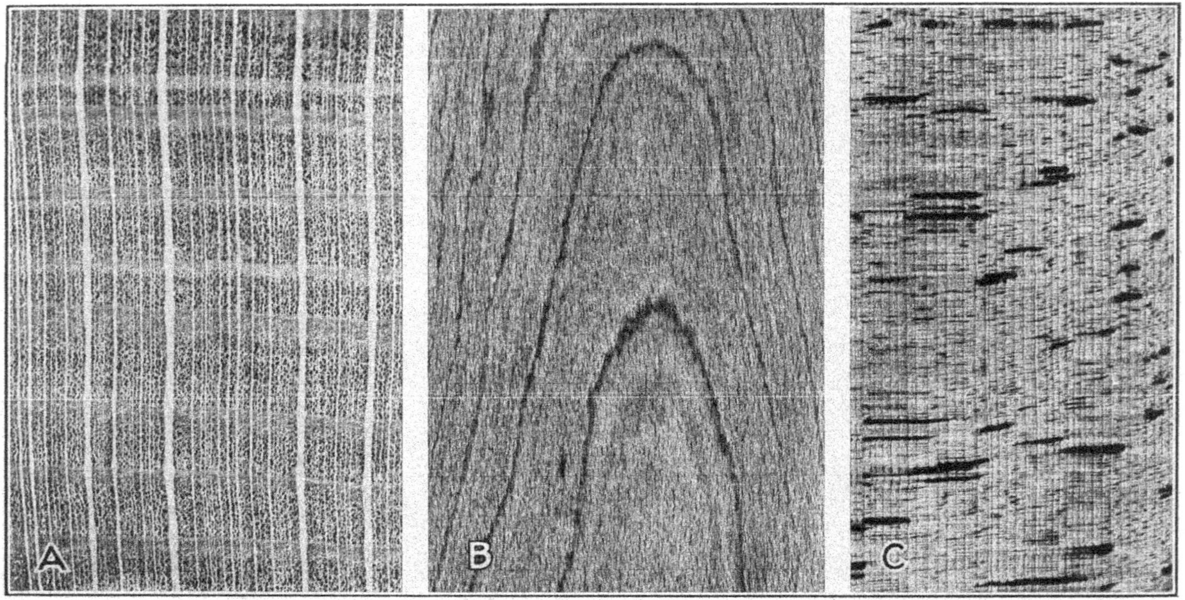

BEECH

A. End surface. (Magnified 7½ diameters.) *B.* Plain-sawed surface. (Natural size.) *C.* Quarter-sawed surface. (Natural size)

It is similar to sugar maple in strength, hardness, and shrinkage, but manifests a greater tendency to warp.

MAPLE

Sugar maple—*Acer saccharum* Marsh.
Red maple—*Acer rubrum* Linn.
Silver maple—*Acer saccharinum* Linn.
Bigleaf maple—*Acer macrophyllum* Pursh.

OTHER NAMES

Sugar maple is commonly called hard maple in distinction from the other two eastern species generally known as soft maples. It is also called rock maple and, in one variety, black maple. Red and silver maple lumber is rarely given any other name but soft maple, although occasionally the former is called swamp maple, water maple, or scarlet maple; the latter white maple, water maple, or river maple. Bigleaf maple is sometimes called broadleaved maple or Oregon maple.

WHERE GROWN

Sugar maple grows throughout the area from Newfoundland to eastern North Dakota and south to the Gulf, except in the peninsula of Florida. Its commercial range is confined to the Lake States, the Northeast, and the Appalachian Mountains to northern Georgia.

Red maple grows in approximately the same territory, but does not grow in the greater part of the State of Minnesota or in adjacent portions of Iowa and Wisconsin.

Silver maple occurs throughout the eastern half of the United States with the exception of the northern parts of Minnesota, Wisconsin, and Michigan, and the extreme southern part of the country.

Bigleaf maple grows along the Pacific coast from San Diego County, Calif., to Alaska. It is used locally for furniture manufacture.

COLOR AND FIGURE

The heartwood of these four maples is light reddish brown; the wide sapwood is practically white when fresh, but may stain to a pale brown if improperly seasoned. The sapwood is often used separately from the heartwood, in which case it is known as "white maple." Sugar maple has more luster than either of the other two.

Straight-grained maple has no figure of commercial value.

Bird's-eye figure, already described on page 19, occurs in some maple trees. It is illustrated in Plate 9, *B*.

Curly figure is occasionally found, sometimes in connection with bird'e-eye grain. When the curly figure assumes fantastical designs, it is sometimes spoken of as landscape figure. Wavy figures, including fiddle-back, also occur in maple, but are not common.

STRUCTURE

End surfaces.—The annual rings are distinctly defined by thin reddish-brown layers. The pores are invisible without a lens.

In sugar maple some of the rays are very small (not visible without a lens) and others comparatively large (very distinct without a lens), whereas in the two soft maples they are more uniform in size and appear to be more numerous. Compare Plate 25, *A*, with Plate 26, *A*.

Longitudinal surfaces.—On plain-sawed surfaces the thin reddish-brown layers defining the annual rings are plainly visible. The medullary rays are visible as exceedingly fine dashes running parallel with the grain. On quarter-sawed surfaces the rays, though small (less than one-sixteenth of an inch high) are very conspicuous as reddish-brown flakes, especially in sugar maple. The pores are not visible on longitudinal surfaces (*B* and *C* of pls. 25 and 26).

The soft maples are very susceptible to attack by the larvæ of an insect which produces "pith flecks"—brown streaks running more or less parallel with the grain. These streaks are the grub holes, closed by a growth of dark-colored cells.[13] Plate 26, *A*, shows two of them in a cross section of red maple; and Figures 2 and 3 of the same plate show them in tangential and radial surfaces, where they appear rather short because they run into the wood at a slight angle.

Pith flecks are not always present in the wood of the soft maples, and they occur occasionally in sugar maple. Hence their presence or absence can not be used as a sole criterion in distinguishing the maples; but whenever pith flecks are abundant there is good reason for suspecting that the wood is one of the soft maples.

USES IN FURNITURE

Sugar maple is one of the pricipal American furniture woods. When used for exposed parts it usually is given a "natural" finish, for which reason wood with curly or bird's-eye figure is more desired. With the "natural" finish its principal use is for bedroom, porch, and kitchen furniture. Occasionally it is used in combination with other woods for exposed parts which are stained or painted and for interior parts where strength or rigidity are essential.

Most maple furniture is made of the solid wood, although highly figured veneer is used to some extent for exposed panels and veneer of a lower grade for backing of bodies and mirrors.

The soft maples are relatively little used in the furniture trade, but occasional pieces are worked in with sugar maple.

Sugar maple ranks with yellow birch and black walnut in certain characteristics which adapt it admirably to furniture construction. These are (1) a high degree of strength and hardness (but not so great hardness as to dull tools excessively), (2) ability to stay in place, and (3) good gluing properties.

BLACK CHERRY

Prunus serotina Ehrh.

OTHER NAMES

This species is known also as wild black cherry, wild cherry, and rum cherry.

WHERE GROWN

Black cherry occurs throughout the United States east of the 97th meridian (except in southern Florida) and in the southern part of eastern Canada, but it is nowhere abundant.

[13] BROWN, H. P., PITH-RAY FLECKS IN WOOD. For. Serv. Cir. 215. 1913.

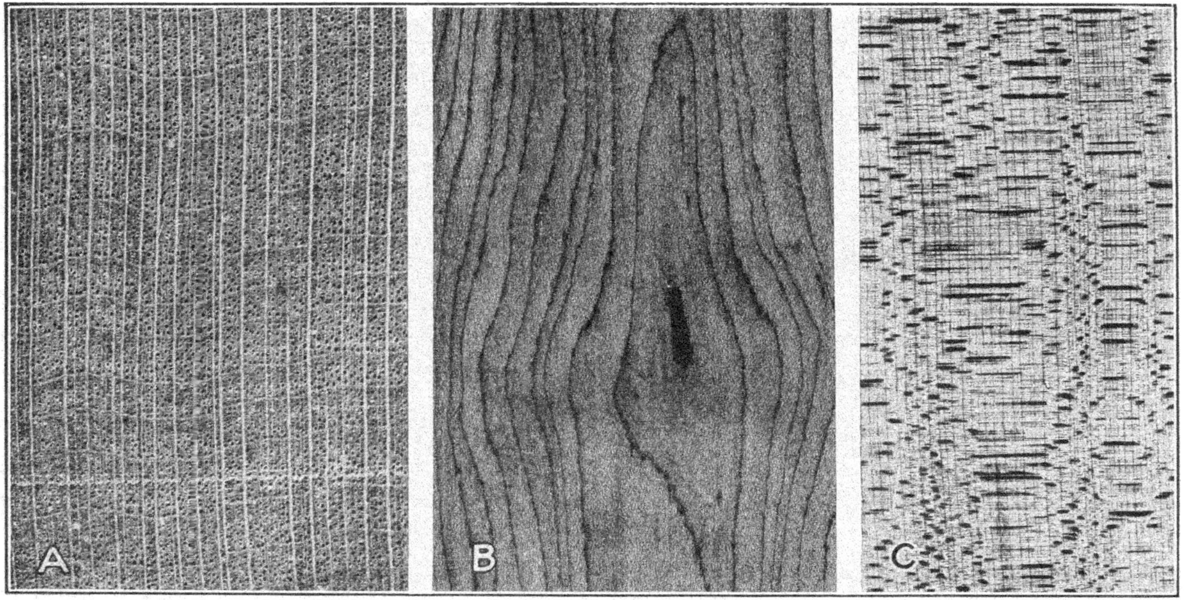

SUGAR MAPLE

A. End surface. (Magnified 7½ diameters.) *B.* Plain-sawed surface. (Natural size.) *C.* Quarter-sawed surface. (Natural size)

101140°—26——5

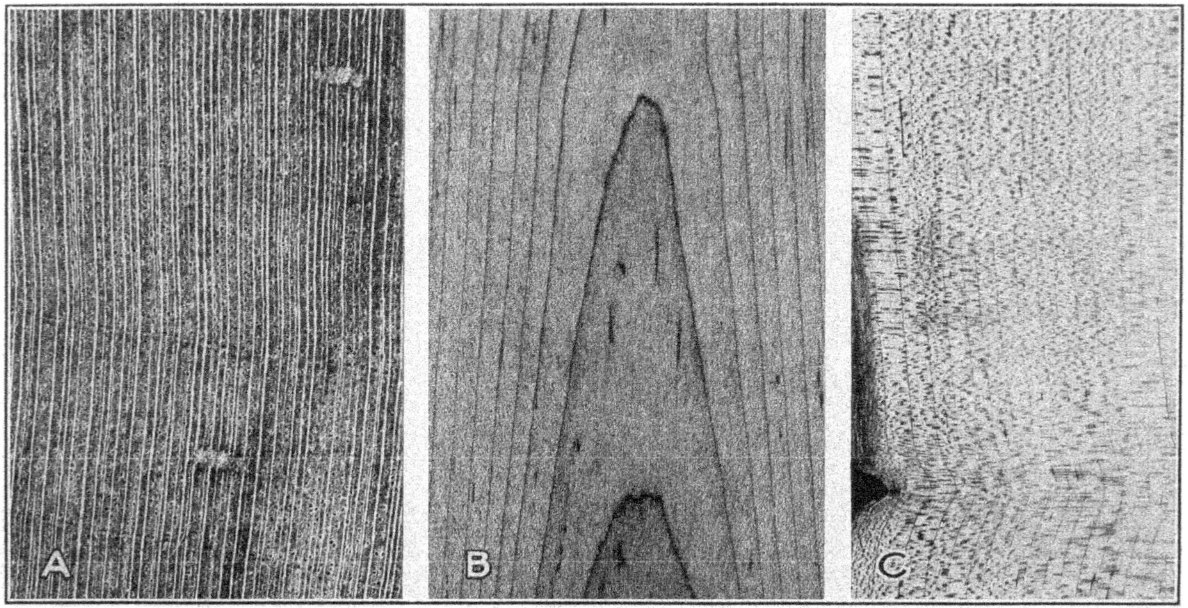

RED MAPLE

A. End surface. (Magnified 7½ diameters.) *B*. Plain-sawed surface. (Natural size.) *C*. Quarter-sawed surface. (Natural size)

COLOR AND FIGURE

The heartwood is from light to dark reddish brown. The narrow sapwood is nearly white. Cherry has no striking figure except in finished pieces cut from burls (pl. 9, *A*), but it posseses a natural luster which adds to its value as a cabinet and furniture wood.

STRUCTURE

End surfaces.—The annual rings are fairly distinct, being defined by an abrupt difference in the size of the pores in adjoining portions.

The pores are small, not visible to the unaided eye or occasionally barely visible; nevertheless, because they decrease in size toward the outer portion of each annual ring the boundaries of the rings are very distinctly marked (pl. 27, *A*).

The rays are very distinct and numerous.

Longitudinal surfaces.—On plain-sawed surfaces the annual growth layers are fairly distinct but not conspicuous. The pores often are visible as very fine grooves. On quarter-sawed surfaces the annual layers are poorly defined; the rays are very distinct but small (pl. 27, *B* and *C*).

USES IN FURNITURE

Cherry, because of its relative scarcity, is not at present used extensively in making furniture, although it has excellent qualities for the purpose. The occasional appearance of articles made of cherry among secondhand furniture, heirlooms, and bric-a-brac indicates that a generation or more ago it was used more than it is now. Cherry resembles unfigured mahogany and is sometimes used in combination with it or as an imitation.

RED GUM

Liquidambar styraciflua Linn.

OTHER NAMES

This species is very commonly known as sweet gum; also as star-leaved gum, hazlewood, and satin walnut (Europe). Commercially the name "red gum" applies to the heartwood only, the sapwood being marketed as "sap gum."

WHERE GROWN

Red gum occurs, in the United States, south of a line from Providence, R. I., through southern Illinois and eastern Texas (except in southern Florida); and in the mountains and highlands of central and southern Mexico and Guatemala. Its commercial range is restricted, however, to the moist lands of the lower Ohio and Mississippi basins and the southeastern and Gulf coasts.

COLOR AND FIGURE

The heartwood is reddish brown. Much red gum lumber contains irregular dark streaks due to natural deposits of coloring

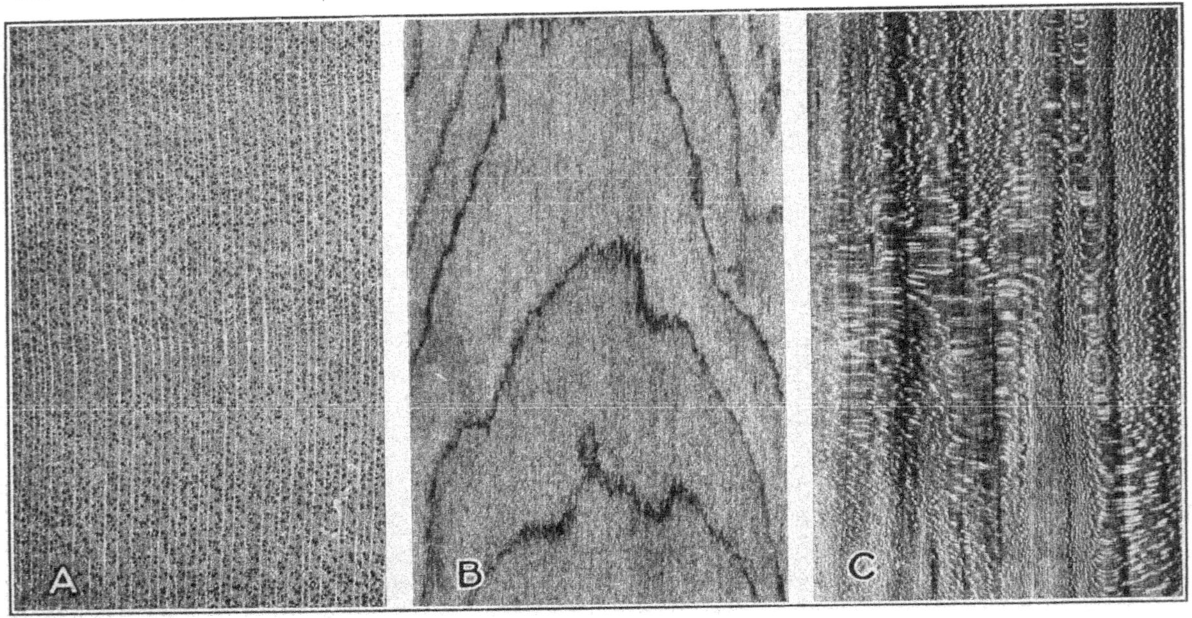

BLACK CHERRY

A. End surface. (Magnified 7½ diameters.) *B*. Plain-sawed surface. (Natural size.) *C*. Quarter-sawed surface. (Natural size)

matter in the wood. Such lumber is called "figured red gum" and that without such streaks "plain gum." The sapwood is pinkish white, unless blued by sap-stain fungi or darkened to a pinkish red by steaming.

The figure is due almost entirely to the irregular dark streaks, which resemble those in Circassian walnut. It is equally pronounced in plain-sawed and quarter-sawed lumber. Some striking patterns can be made from figured red gum veneer when properly matched, as for instance those shown in Plate 4, *B*.

Occasionally quarter-sawed boards exhibit a stripe figure due to interlocked grain, but because the wood is not very lustrous the stripe is not so noticeable as in some other woods.

STRUCTURE

End surfaces.—Although annual rings are present, they are not very distinct even under a lens.

The pores are so small as to be invisible without magnification. Under a lens they appear approximately uniform in size and evenly distributed.

The rays are fairly distinct without magnification and are comparatively numerous (pl. 28, *A*).

Longitudinal surfaces.—The annual rings and sections through pores are not distinct on plain or quarter-sawed surfaces.

On strictly radially cut surfaces the medullary rays are readily visible but not conspicuous.

USES IN FURNITURE

Red gum is one of the leading furniture woods of the country. For the better grades of furniture the heartwood is preferred, but the sapwood ("sap gum") also is used extensively for panel cores and even exteriors.

Unmixed with other woods this species is used in the manufacture of "gum" furniture; but more often it is used in combination with walnut or mahognay, both of which it can be made to resemble by careful staining, so that it requires considerable familiarity with the woods to say offhand which is which. Both walnut and mahogany, however, are readily distinguishable under good illumination by their larger pores, visible to the unaided eye.

Red gum heartwood not only lends itself easily to various finishes but often has a beautiful figure of its own and so deserves recognition under its proper name. The sapwood, although nearly white, also is easily stained in various shades. Improved seasoning methods and the use of veneers on cross-banded panels have practically done away with the difficulties experienced when this species was first used for high-grade work. It is softer and weaker than walnut or yellow birch, but still within the range which gives good service under ordinary conditions. The ease with which it can be worked on machines and its adaptability to various finishes are the chief qualities which have made red gum prominent in the furniture trade.

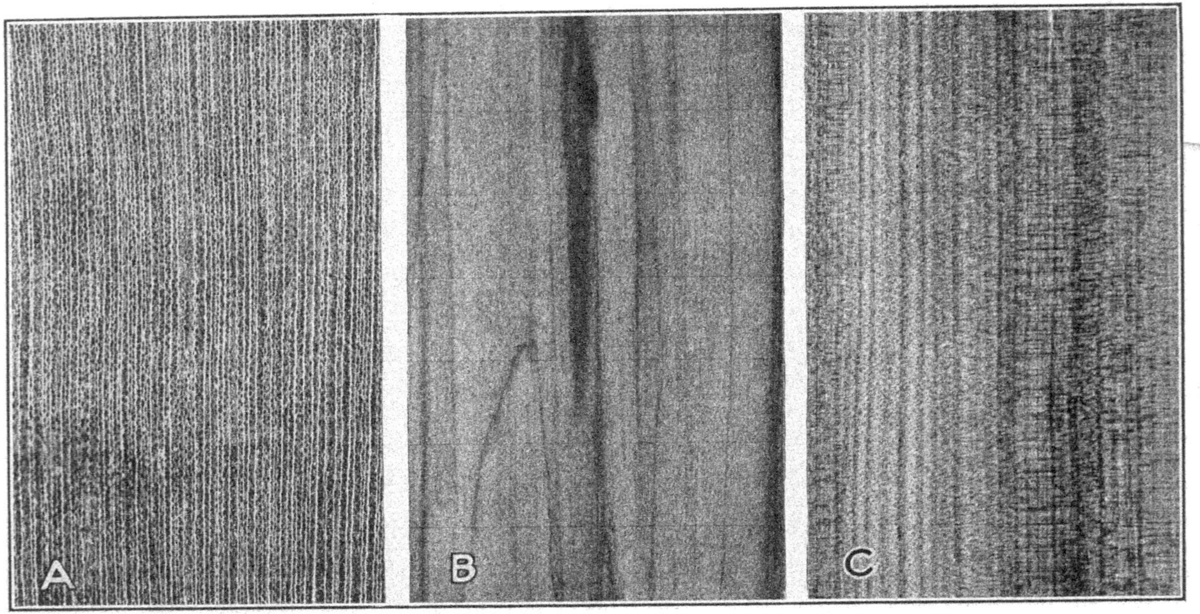

RED GUM

A. End surface. (Magnified 7½ diameters.) *B.* Plain-sawed surface. (Natural size.) *C.* Quarter-sawed surface. (Natural size)

TUPELO GUM

Nyssa aquatica Linn.

OTHER NAMES

Cotton gum, tupelo, swamp tupelo, and bay poplar are other names applied to this species.

WHERE GROWN

Tupelo gum occurs throughout the coast region from southern Virginia to northern Florida, and through the Gulf States to Texas; and northward through the Mississippi River bottom lands to southern Illinois.

COLOR AND FIGURE

The white sapwood merges gradually into the pale brownish-gray heartwood. Although tupelo gum has interlocking grain as a rule, it lacks the luster necessary to bring out the stripe or ribbon effect to full advantage on radially cut surfaces. No other types of figure are commonly found in this species.

STRUCTURE

End surfaces.—Although the annual rings are well defined by narrow lines, they are not conspicuous even under a lens. The pores are too small to be seen with the unaided eye. Their even distribution gives the wood a homogeneous appearance. The rays are not visible without a lens (pl. 29, *A*).

Longitudinal surfaces.—Cut lengthwise the annual rings and rays are inconspicuous and the pores are faintly visible. A characteristic of tupelo gum as seen by the unaided eye appears to be the absence rather than the presence of distinctive colors or structural features.

USES IN FURNITURE

Tupelo gum has recently been used in considerable quantity for cores of veneered panels and even for exposed parts in place of red gum, from which it can best be distinguished by its lack of pink or reddish tints.

The heavier grades of tupelo gum can not readily be distinguished from black gum (*Nyssa sylvatica* Marsh.). The latter is very similar in properties and is used to a small extent by the furniture trade. Much of the tupelo gum, however, is lighter in weight and warps less easily, and is therefore preferred by the manufacturers. On the other hand, quartered black gum veneer has lately gained prominence in the furniture industry, because of its stripe figure due to interlocked grain.

In general, tupelo gum is similar to red gum in strength and hardness; but the lighter-weight lumber is weaker and softer.

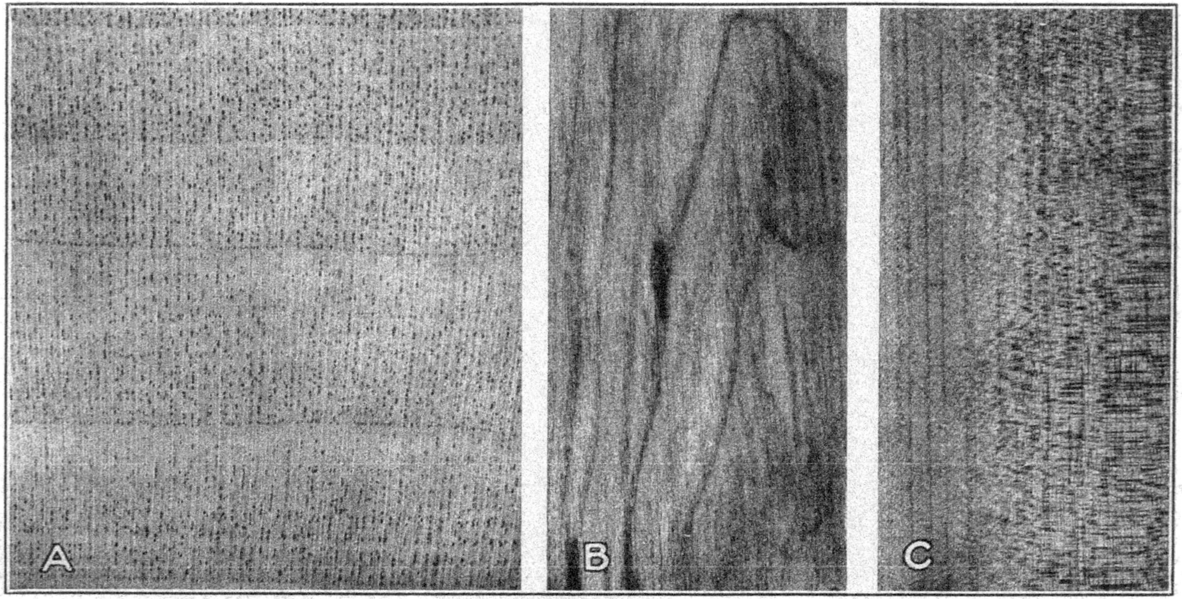

TUPELO GUM

A. End surface. (Magnified 7½ diameters.) *B*. Plain-sawed surface. (Natural size.) *C*. Quarter-sawed surface. (Natural size)

RED ALDER

Alnus rubra Bong.

OTHER NAMES

Western alder, alder.

WHERE GROWN

Pacific coast from Santa Barbara, Calif., to Sitka, Alaska.

COLOR AND FIGURE

The heartwood and sapwood are of about the same color, ranging from pale pinkish brown to almost white. Green lumber when exposed to the air turns reddish brown at the surface, but the discoloration dresses off in planing.

The annual rings produce an obscure pattern on plain-sawed surfaces, but otherwise the wood forms no ornamental figure.

STRUCTURE

End surfaces.—Annual layers of growth can be distinguished, but they are not conspicuous. The pores are not visible without a lens. Some boards contain large rays as wide and distinct as in oak, ranging in spacing from less than one-eighth of an inch to over 1 inch apart, whereas others have only small rays which are not visible without a lens (pl. 30, *A*).

Longitudinal surfaces.—The annual rings are plainly visible on both tangential and radial surfaces. The pores, although even smaller than in birch, are visible as minute grooves. When large rays are present they appear on tangential surfaces as darker, heavy lines extending from one-fourth to 2 or more inches lengthwise of the board. On radial surfaces the large rays produce a wide flake, as in oak, but the milder contrast between rays and wood fiber renders them less conspicuous. The small rays are inconspicuous (pl. 30, *A* and *B*).

USES IN FURNITURE

Red alder is manufactured on the west coast into cores of panels for table tops, dresser and chiffonier tops, sides, and drawer fronts, etc. It is also used for exposed parts of kitchen furniture, and for legs, mirror frames, and similar solid parts of household furniture in general. The fact that it seasons easily, turns and carves well, holds its shape, and stains readily in imitation of mahogany or walnut makes is particularly desirable for such exposed parts. In hardness and in strength as a beam or post it ranks between red gum and yellow poplar.

KEY FOR THE IDENTIFICATION OF WOODS USED FOR EXPOSED PARTS OF FURNITURE

In the following key the distinguishing features of furniture woods are arranged in such a way that a given piece can be "traced

PLATE XXX

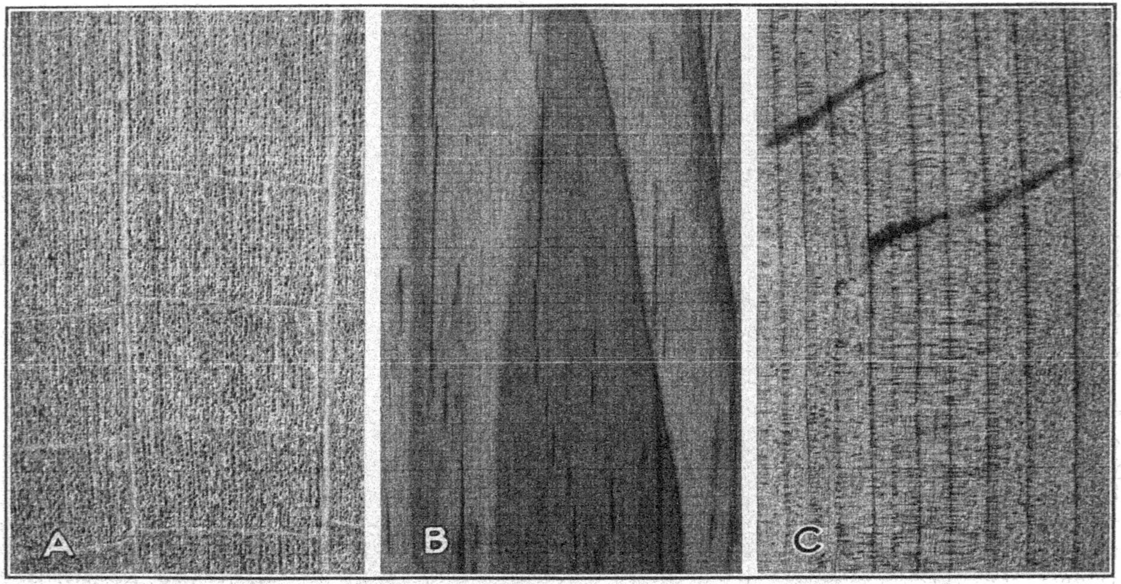

RED ALDER

A. End surface. (Magnified 7½ diameters.) *B*. Plain-sawed surface. (Natural size.) *C*. Quarter-sawed surface. (Natural size)

down" and identified much as a flower is identified by a botanist. In trying to identify a piece of wood by this key the first thing to do is to examine the smoothly cut end surface [14] to see whether or not there is a zone of large pores at the beginning of each annual ring. According to this structural feature the wood can be classified as falling in either Group I or Group II. Let us take, for example, an unknown wood, and say it falls in Group II (p. 75). The next observation to make is whether it can be classed under "A. Pores visible without a magnifying glass," etc., or "B. Pores not visible without a magnifying glass," etc., always bearing in mind that such observations are to be made on the end surfaces, unless otherwise stated. We find, let us say, that the pores are not merely visible but plainly visible. This classifies the specimen under "2. Pores plainly visible on smoothly cut end surfaces." Next we note that the pores contain a dark-colored gum, and the wood is further classified as coming under "AA." Further observations show that the natural color [15] of the heartwood on longitudinal surfaces is a fairly uniform reddish brown. This classifies the wood under "(*b*). Heartwood reddish brown," etc., and it must be either true mahogany or khaya. Another examination of the end surface shows that the growth rings are not clearly defined, and the wood therefore must be khaya.

Difficulties undoubtedly will arise in using the key; for example, as in a finished piece of furniture, it may be undesirable to make a smooth cut across the grain, or the piece to be identified may be thin veneer in which the structural features can not be definitely determined from the end. In such cases it is often possible to get enough information from an observation of the longitudinal surface to identify the wood, especially by referring to the descriptions (pp. 27 to 71) and the illustrations. In case of doubt the Forest Products Laboratory, Madison, Wis., will be glad to assist if a small sliver of the wood is sent for microscopic identification.

The key or system of identification here presented is based principally on the structure of the wood as seen without magnification. Unless otherwise stated, all observations as to structure should be made on smoothly cut (not sanded) end surfaces and all observations as to color on freshly cut longitudinal surfaces of the heartwood. The name follows the description.

RING-POROUS SPECIES

I. Woods with zone of large pores at beginning of each annual ring—called "ring-porous."
   A. Many rays comparatively broad and conspicuous, appearing as distinct lines on end surfaces, as large flakes on quarter-sawed surfaces, and as brownish lines from one-half to 4 inches in height on plain-sawed surfaces. The dense outer half of each annual ring figured with

---

[14] The cut should be made with a very sharp knife. Sandpapering the surface will not do, since it more or less obliterates the structure.

[15] In stained or painted furniture it is often possible to note the natural color on under or interior surfaces.

irregular lines or V-shaped patches extending radially across the rings. Wood heavy to very heavy, and hard [16]--------The oaks, p. 27.

1. The large pores in the heartwood mostly closed with tyloses. Heartwood usually without reddish tinge.[17] White oak group, p. 27.
2. The large pores in the heartwood mostly open. Heartwood usually with reddish tinge [17]------------------------Red oak group, p. 27.

B. All rays comparatively narrow and inconspicuous.

1. The outer half of each annual ring figured with V-shaped *radial* lines or patches, as in oak. Heartwood grayish brown. Wood moderately light in weight and moderately soft----Chestnut, p. 32.
2. The outer half or more of each annual ring figured with more or less wavy lines or narrow bands extending *tangentially* along the rings.

AA. Conspicuous wavy tangential lines or bands throughout entire annual ring except in zone of large pores at beginning of rings-----------------------------------The elms, p. 34.

(*a*) Zone of larger pores at beginning of annual ring consisting of several rows. Heartwood dark reddish brown or chocolate brown in color, sometimes with odor resembling licorice. Sapwood rarely over 1 inch wide. Wood moderately heavy and moderately hard. Slippery elm, p. 34.

(*b*) Zone of larger pores at beginning of annual ring consisting of one row, except in wide rings, which may show more than one row. Heartwood grayish brown or reddish brown but not so dark as slippery elm and without licorice odor. Sapwood usually over 1 inch wide.

(*aa*) Wood heavy and hard. Pores at beginning of each annual ring inconspicuous because comparatively small, not close together, and plugged up more or less with tyloses---------------Rock elm, p. 34.

(*bb*) Wood moderately heavy and moderately hard. Pores at beginning of each annual ring fairly conspicuous because larger and closer together than in rock elm, and open---------American elm, p. 34.

BB. Wavy tangential lines in each annual ring comparatively inconspicuous, and confined mostly to the outer part of the ring.[18] Heartwood light grayish brown. Sapwood over 1 inch, usually several inches wide. Wood heavy and hard. White ash, p. 35. Green ash, p. 35.

---

[16] The following scales of weight and hardness are used in describing woods:

Weight	Species, for example	Hardness
Extremely light		Extremely soft.
Very, very light		Very, very soft.
Very light		Very soft.
Light	Basswood	Soft.
Moderately light	Yellow poplar	Moderately soft.
Moderately heavy	Black cherry	Moderately hard
Heavy	Yellow birch	Hard.
Very heavy	True hickories	Very hard.
Very, very heavy		Very, very hard.
Extremely heavy		Extremely hard.

[17] See p. 29 for a more reliable method of distinguishing between the white and red oak groups with the aid of a magnifying glass.

[18] With a magnifying glass it can be seen that in white ash and green ash the small pores in the outer part of each annual ring are isolated or in groups of two, somewhat scattered, but connected by light-colored tissue; whereas in the elms the pores are very numerous and joined to each other in wavy tangential lines. Compare Plates 13, *A*, 14, *A*, and 15, *A*.

3. No distinct radial or tangential lines or figures visible without a lens within the annual rings.
   - AA. Heartwood reddish brown, large pores at beginning of annual ring not crowded, decreasing in size more or less gradually toward outer limit of each ring. Sapwood over 1 inch wide. Wood very heavy and very hard.
     Pecan (hickory), p. 38.
   - BB. Heartwood decidedly grayish grown. Large pores at beginning of annual ring crowded and decreasing in size abruptly toward outer edge of each ring. Sapwood usually less than 1 inch wide. Wood moderately heavy and moderately hard---------------------------**Black ash, p. 35.**

DIFFUSE-POROUS SPECIES

II. Woods without definite zone of large pores at beginning of each annual ring, although the pores may gradually decrease in size from the inner to the outer limit of each year's growth—called "diffuse-porous."

A. Pores visible without a magnifying glass on smoothly cut end surfaces, appearing as minute holes. Also visible as fine grooves or lines on longitudinal surfaces.

1. Pores barely visible on smoothly cut end surfaces in good light. Heartwood reddish brown. Wood heavy and hard.
   Yellow birch, p. 41.
   Sweet birch, p. 41.
2. Pores plainly visible on smoothly cut end surfaces.
   - AA. Pores contain dark-colored gum seen best on smooth longitudinal surfaces.
     - (*a*) Heartwood in various shades of reddish brown or purplish brown, figured with black streaks. Wood ranks relatively high in weight (some pieces sink in water when dry) and in hardness-----------**Rosewood, p. 43.**
     - (*b*) Heartwood reddish brown in various shades, but each piece fairly uniform in color, except for differences in the reflection of light. Wood moderately light to heavy and moderately soft to hard.
       - (*aa*) Growth rings defined by light-colored lines.
         True mahogany, p. 44.
       - (*bb*) Growth rings not clearly defined-------**Khaya, p. 48.**
   - BB. Pores do not contain dark-colored gum, but contain glistening tyloses, which, however, are hardly visible without a lens.
     - (*a*) Heartwood reddish brown.
       - (*aa*) Annual rings present. Pores decrease in size gradually from inner to outer edge of rings. Wood very heavy and very hard---------**Water hickory, p. 38.**
       - (*bb*) Annual rings absent, although white tangential lines of varying lengths and from one-eighth to several inches apart are present. Pores fairly uniform in size. Wood moderately light to heavy and moderately soft to hard.
         Tanguile, red lauaan, and almon, p. 49.
     - (*b*) Heartwood essentially brown without reddish tinge.
       - (*aa*) Color chocolate brown, occasionally with lavender tinge. Wood heavy and hard---**Black walnut, p. 51.**
       - (*bb*) Color fawn brown, usually interspersed with darker streaks. Wood heavy and hard.
         Circassian walnut, p. 53.

B. Pores not visible without a magnifying glass on end surfaces.

1. Rays comparatively broad on end surfaces, appearing as conspicuous flakes on strictly quarter-sawed surfaces and as dashes on plain-sawed surfaces.
   - AA. Large rays from one-fourth to over 2 inches high (as measured along the grain); and on end surfaces from less than one-eighth to over 1 inch apart. No distinct band of summerwood present. Wood moderately heavy and moderately hard-------------------------------------**Red alder, p. 71.**

Spotted something that isn't quite right?

Let us know and we'll fix it for future reprints. And as a thank you for your diligence, we'll send you a free Sothis book of your choice.

publisher@sothis.press

This page intentionally left blank

Except of course for that line above. And this one.
And whilst we have your attention, perhaps you'd
like to follow us on Instagram @sothis.press. Go on.

Made in the USA
Monee, IL
08 November 2026

68106070R00049